T0214938

SpringerBriefs in Mathematical Physics

Volume 46

SpringerBriefs are characterized in general by their size (50–125 pages) and fast production time (2–3 months compared to 6 months for a monograph).

Briefs are available in print but are intended as a primarily electronic publication to be included in Springer's e-book package.

Typical works might include:

- An extended survey of a field
- A link between new research papers published in journal articles
- A presentation of core concepts that doctoral students must understand in order to make independent contributions
- Lecture notes making a specialist topic accessible for non-specialist readers.

SpringerBriefs in Mathematical Physics showcase, in a compact format, topics of current relevance in the field of mathematical physics. Published titles will encompass all areas of theoretical and mathematical physics. This series is intended for mathematicians, physicists, and other scientists, as well as doctoral students in related areas.

Editorial Board

- Nathanaël Berestycki (University of Cambridge, UK)
- Mihalis Dafermos (University of Cambridge, UK / Princeton University, US)
- Atsuo Kuniba (University of Tokyo, Japan)
- Matilde Marcolli (CALTECH, US)
- Bruno Nachtergaele (UC Davis, US)
- Hal Tasaki (Gakushuin University, Japan)
- 50 – 125 published pages, including all tables, figures, and references
- Softcover binding
- Copyright to remain in author's name
- Versions in print, eBook, and MyCopy

Asao Arai

Infinite-Dimensional Dirac Operators and Supersymmetric Quantum Fields

An Introduction to Analysis on Boson–Fermion Fock Spaces

 Springer

Asao Arai
Department of Mathematics
Hokkaido University
Sapporo, Hokkaido, Japan

ISSN 2197-1757   ISSN 2197-1765   (electronic)
SpringerBriefs in Mathematical Physics
ISBN 978-981-19-5677-5   ISBN 978-981-19-5678-2   (eBook)
https://doi.org/10.1007/978-981-19-5678-2

This Springer imprint is published by the registered company Springer Nature Singapore Pte Ltd.
The registered company address is: 152 Beach Road, #21-01/04 Gateway East, Singapore 189721,
Singapore

Preface

This book originates from my work on infinite-dimensional analysis in relation to mathematical studies of supersymmetric quantum field theory, which started in 1984 and still continues. During the course of the research, I discovered an intrinsic mathematical structure that some models in supersymmetric quantum field theory have in common. It may be characterized as the theory of *infinite-dimensional Dirac operators on the abstract boson–fermion Fock space*, the tensor product Hilbert space of the abstract boson Fock space and the abstract fermion Fock space. I consider it one of the most important results in this research and to be developed further.

The present book is written as an introduction to analysis of the abstract boson–fermion Fock space with applications to mathematical supersymmetric quantum field theory, which forms an interesting field of infinite-dimensional analysis. The emphasis is put on the theory of infinite-dimensional Dirac operators as suggested above. Since infinite-dimensional Dirac operators may have relations to infinite-dimensional geometry, the book may be read also from that viewpoint.

A general background behind the infinite-dimensional analysis treated in the book is found in the abstract supersymmetric quantum mechanics (SQM). For this reason, we begin with a review of the mathematical theory of it in Chap. 1. The abstract SQM can be applied to both SQM with finite degrees of freedom and SQM with infinite degrees of freedom including supersymmetric quantum field theory. In Chap. 2, we summarize fundamental aspects of the theory of Fock spaces (full Fock space, boson Fock space and fermion Fock space) within the scope of the following chapters. In Chap. 3, we review the Q-space representation—a probability-theoretical representation—of the abstract boson Fock space, which is useful to derive path (functional) integral representations for vacuum expectation values or traces of operators with respect to the heat semi-groups generated by boson second quantization operators and their perturbations. Chapter 4 is the main body of the present book and is devoted to an introductory description of the theory of infinite-dimensional Dirac operators on the abstract boson-fermion Fock space. We see that the theory is a realization of infinite Hilbert complexes as general concepts. Moreover, we construct an abstract interacting supersymmetric quantum field model in terms of infinite-dimensional Dirac operators on the abstract boson–fermion Fock space. In the last chapter, we show that the two-dimensional $N = 1$ and $N = 2$ Wess–Zumino models in supersymmetric quantum field theory are concrete realizations of the abstract supersymmetric quantum field model introduced in Chap. 4 and hence that the theory in Chap. 4 gives an abstract unification of the models.

Since the book is introductory as mentioned above, two appendices are added. In Appendix A, self-adjoint extensions of a symmetric operator matrix are described. Appendix B concerns the construction of an infinite-dimensional Gaussian measure on the space of continuous functions with values in a real Hilbert space on a finite interval (a path space). For the same reason, the references are not intended to be complete.

The intended audience for the present book is mainly graduate students and non-experts in mathematics and mathematical physics who are interested in infinite-dimensional analysis as well as mathematical analysis of quantum field theories, including supersymmetric ones.

I would like to thank Roman Gielerak for inviting me to deliver a series of lectures on infinite-dimensional Dirac operators at the XXVIII Karpacz Winter School of Theoretical Physics, Poland, 1992, Rémi Léandre and Sylvie Paycha for their kind interest in my work and inviting me to Institut de Recherche Mathématique Avancée (IRMA), Université de Strasbourg in 1994 and Itaru Mitoma (deceased) for joint work on infinite-dimensional analysis on the abstract boson–fermion Fock space. My thanks go also to Mr. Masayuki Nakamura, editor, at Springer Japan for inviting me to write a book in the series of Springer Briefs in Mathematical Physics.

Sapporo, Japan   Asao Arai
June 2022

Contents

Symbols

Numbers

$\mathbb{N}$	Natural numbers
$\mathbb{Z}_+$	Non-negative integers
$\mathbb{Z}$	Integers
$\mathbb{R}$	Real numbers
$\mathbb{C}$	Complex numbers
i	Imaginary unit
Im z	Imaginary part of $z \in \mathbb{C}$

Operational Symbols

$\{\,,\,\}$	Anti-commutator: $\{A, B\} := AB + BA$ (A and B denote linear operators)[1]
$[\,,\,]$	Commutator: $[A, B] := AB - BA$
$\langle\,,\,\rangle_{\mathcal{H}}$	Inner product of Hilbert space $\mathcal{H}$ (linear in the second vector)
$\langle\,,\,\rangle$	Inner product of Hilbert space
$\|\cdot\|_{\mathcal{H}}$	Norm of $\mathcal{H}$
$\|\cdot\|$	Norm of Hilbert space
$\|\cdot\|_2$	Hilbert–Schmidt norm
$\widehat{\otimes}^n \mathscr{D}$	n-fold algebraic tensor product of a vector space $\mathscr{D}$
$\widehat{\wedge}^n \mathscr{D}$	n-fold algebraic anti-symmetric tensor product of $\mathscr{D}$
$\widehat{\otimes}_s^n \mathscr{H}$	n-fold algebraic symmetric tensor product of $\mathscr{D}$
$\otimes^n \mathscr{H}$	n-fold tensor product Hilbert space of $\mathscr{H}$
$\wedge^n(\mathscr{H})$	n-fold anti-symmetric tensor product Hilbert space of $\mathscr{H}$
$\otimes_s^n \mathscr{H}$	n-fold symmetric tensor product Hilbert space of $\mathscr{H}$
$A \otimes B$	Tensor product of A and B

---

[1] "$X := Y$" indicates that X is defined by Y.

$A \widehat{\otimes} B$	Algebraic tensor product of A and B
$\oplus_{n=0}^{\infty} \mathcal{H}_n$	Infinite direct sum of Hilbert spaces $\mathcal{H}_n$, $n \geq 0$
$\widehat{\oplus}_{n=0}^{\infty} \mathcal{D}_n$	Algebraic infinite direct sum of vector spaces $\mathcal{D}_n$, $n \geq 0$
span $\mathcal{D}$	Subspace algebraically spanned by a subset $\mathcal{D}$ of a vector space

Other Symbols

$A \subset B$	Operator B is an extension of operator A
	(A is a restriction of B)
$\overline{A}$	Closure of a closable linear operator A
A^*	Adjoint of a linear operator A when A is densely defined
$A \upharpoonright \mathcal{D}$	Restriction of A to a subspace $\mathcal{D}$
$\|A\|$	Modulus of A when A is densely defined and closed
$A(f)$	Boson annihilation operator with test vector f
$A(f)^*$	Boson creation operator with test vector f
$\mathfrak{B}(\mathcal{H})$	Bounded linear operators everywhere defined on $\mathcal{H}$
$B(u)$	Fermion annihilation operator with test vector u
$B(u)^*$	Fermion creation operator with test vector u
$\mathfrak{C}(\mathcal{H}, \mathcal{K})$	Densely defined closed linear operators from $\mathcal{H}$ to $\mathcal{K}$ (Hilbert space)
CONS	Complete orthonormal system
$\overline{\mathcal{D}}$	Closure of a subset $\mathcal{D}$
$\mathcal{D}^{\perp}$	Orthogonal complement of a subset $\mathcal{D}$ of a Hilbert space
δ_{ab}	Kronecker delta: $a = b \Rightarrow \delta_{ab} = 1$; $a \neq b \Rightarrow \delta_{ab} = 0$
Δ_S	Laplace–Beltrami operator associated with $S \in \mathfrak{C}(\mathcal{H}, \mathcal{K})$
$\Delta_{S,p}$	p-th Laplace–Beltrami operator associated with $S \in \mathfrak{C}(\mathcal{H}, \mathcal{K})$
$\det(1 + T)$	Determinant of $1 + T$ when T is trace class
$\det_2(1 + T)$	Regularized determinant of T when T is Hilbert–Schmidt
D_f	Directional functional differential operator
$d\Gamma(T)$	Second quantization operator of a densely defined closed linear operator T
$d\Gamma_{\mathrm{b}}(T)$	Boson second quantization operator of T
$d\Gamma_{\mathrm{f}}(T)$	Fermion second quantization operator of T
$\dim \mathcal{V}$	Dimension of a vector space $\mathcal{V}$
$\mathrm{Dom}(A)$	Domain of a linear operator A
d_S	Exterior differential operator associated with $S \in \mathfrak{C}(\mathcal{H}, \mathcal{K})$
E_A	Spectral measure of A when A is self-adjoint
$\mathscr{F}(\mathcal{H})$	Full Fock space over $\mathcal{H}$
$\mathscr{F}_{\mathrm{b}}(\mathcal{H})$	Boson Fock space over $\mathcal{H}$
$\mathscr{F}_{\mathrm{f}}(\mathcal{H})$	Fermion Fock space over $\mathcal{H}$
$\mathscr{F}(\mathcal{H}, \mathcal{K})$	Boson–fermion Fock space over $(\mathcal{H}, \mathcal{K})$
$\Gamma(T)$	Γ-operator of T
$\Gamma_{\mathrm{b}}(T)$	Boson Γ-operator of T

$\Gamma_{\mathrm{f}}(T)$	Fermion Γ-operator of T
I	Identity
$\mathrm{ind}\,A$	index of A
$\ker A$	kernel of A
∇	Gradient operator on the abstract boson Fock space in the Q-space representation
N_{b}	Boson number operator
N_{f}	Fermion number operator
N_{tot}	Total number operator
$\mathrm{nul}(A)$	Nullity of A: $\mathrm{nul}(A) := \dim \ker A$
$\mathscr{P}_n$	Complex polynomials of n variables
Q_S	Dirac operator associated with $S \in \mathfrak{C}(\mathscr{H}, \mathscr{K})$
$\mathrm{Ran}(A)$	Range of A
$\sigma(A)$	Spectrum of A
$\sigma_{\mathrm{p}}(A)$	Point spectrum of A
$\mathrm{Tr}\,A$	trace of A when A is trace class
$\varphi_{\mathfrak{h}}(\cdot)$	Gaussian random process indexed by a real Hilbert space $\mathfrak{h}$
$\Omega_{\mathscr{H}}$	Fock vacuum in $\mathscr{F}(\mathscr{H})$

Chapter 1
Abstract Supersymmetric Quantum Mechanics

Abstract We review basic aspects of the mathematical theory of abstract super-symmetric quantum mechanics, which can be applied to supersymmetric quantum mechanics both with finite degrees of freedom and with infinite degrees of freedom, including supersymmetric quantum field theory.

1.1 Definition and Basic Properties

An abstract form of supersymmetric quantum mechanics is defined as follows [5, 64] (for physical backgrounds, see, e.g., [66–68]):

Definition 1.1 Let $N \in \mathbb{N}$. A quadruple $(\mathscr{H}, \Gamma, \{Q_i\}_{i=1}^N, H)$ consisting of a Hilbert space[1] $\mathscr{H}$ and linear operators Γ, Q_i $(i = 1, \ldots, N)$, H on $\mathscr{H}$ satisfying the following (i)–(iv) is called a **supersymmetric quantum mechanics** (SQM):

(i) Γ is self-adjoint and unitary with $\Gamma \neq \pm I$ (I denotes identity).

(ii) For each $i = 1, \ldots, N$, Q_i is self-adjoint and $H = Q_i^2$.

(iii) For each $i = 1, \ldots, N$, Γ leaves $\mathrm{Dom}(Q_i)$, the domain of Q_i, invariant (i.e. $\Gamma(\mathrm{Dom}(Q_i)) \subset \mathrm{Dom}(Q_i)$) and anti-commutes with Q_i on $\mathrm{Dom}(Q_i)$: for all $\Psi \in \mathrm{Dom}(Q_i)$,

$$\{\Gamma, Q_i\}\Psi = 0 \ , \tag{1.1}$$

where $\{,\}$ denotes anti-commutator: $\{A, B\} := AB + BA$ for algebraic objects A and B.

(iv) In the case $N \geq 2$, for all $i, j = 1, \ldots, N$ with $i \neq j$, Q_i and Q_j anti-commute on $\mathrm{Dom}(Q_i) \cap \mathrm{Dom}(Q_j)$ in the sense of sesquilinear form:

---

[1] In this book, we mean by a "Hilbert space" a complex Hilbert space unless otherwise stated. We denote the inner product and the norm of a Hilbert space $\mathscr{H}$ by $\langle , \rangle_{\mathscr{H}}$ (linear in the second vector) and $\| \cdot \|_{\mathscr{H}}$ respectively. But, if there is no danger of confusion, we write them simply $\langle , \rangle$ and $\| \cdot \|$ respectively.

© The Author(s), under exclusive license to Springer Nature Singapore Pte Ltd. 2022   1
A. Arai, *Infinite-Dimensional Dirac Operators and Supersymmetric Quantum Fields*,
SpringerBriefs in Mathematical Physics,
https://doi.org/10.1007/978-981-19-5678-2_1

$$\langle Q_i \Psi, Q_j \Phi \rangle + \langle Q_j \Psi, Q_i \Phi \rangle = 0, \quad \Psi, \Phi \in \mathrm{Dom}(Q_i) \cap \mathrm{Dom}(Q_j).$$

In Definition 1.1, $\mathscr{H}$ denotes the Hilbert space of state vectors of the SQM. It follows from condition (ii) that H is a non-negative self-adjoint operator on $\mathscr{H}$. Each operator Q_i and H are called a **self-adjoint supercharge** (or simply a supercharge) and the **supersymmetric Hamiltonian** in the SQM respectively. The number N of self-adjoint supercharges is called the **degree** of supersymmetry. If $N = n$ for a specific natural number n, then $(\mathscr{H}, \Gamma, \{Q_i\}_{i=1}^n, H)$ is called an $N = n$ SQM. We denote an $N = 1$ SQM by $(\mathscr{H}, \Gamma, Q, H)$ $(Q := Q_1)$.

For a linear operator A on a Hilbert space, we denote by $\sigma(A)$ (resp. $\sigma_{\mathrm{p}}(A)$) the spectrum (resp. the point spectrum) of A.[2]

Condition (i) in Definition 1.1 implies that

$$\Gamma^2 = I, \quad \sigma(\Gamma) = \sigma_{\mathrm{p}}(\Gamma) = \{-1, 1\}, \tag{1.2}$$

hence Γ is a grading operator on $\mathscr{H}$.[3] It follows from condition (ii) that

$$\mathrm{Dom}(H^{1/2}) = \mathrm{Dom}(|Q_i|) = \mathrm{Dom}(Q_i) \quad (i = 1, \ldots, N), \quad |Q_i| = H^{1/2}, \tag{1.3}$$

where, for a self-adjoint operator A, $|A|$ denotes the modulus of A: $|A| := \int |\lambda| dE_A(\lambda)$ (E_A is the spectral measure of A) and, if A is non-negative, then $A^{1/2}$ is defined by $A^{1/2} := \int \lambda^{1/2} dE_A(\lambda)$.[4]

1.2   Reflection Symmetry of the Spectrum of a Self-adjoint Supercharge

Let $(\mathscr{H}, \Gamma, \{Q_i\}_{i=1}^N, H)$ be an SQM and denote any Q_i by Q. Then:

Theorem 1.1

(i) *The spectrum $\sigma(Q)$ is reflection symmetric with respect to the origin of $\mathbb{R}$, i.e., if $\lambda \in \sigma(Q)$, then $-\lambda \in \sigma(Q)$.*

---

[2] $\sigma(A) := \mathbb{C} \setminus \rho(A)$, where $\rho(A) := \{z \in \mathbb{C} | A - z$ is injective and $\mathrm{Ran}(A - z)$, the range of $A - z$, is dense with $(A - z)^{-1}$ being bounded$\}$, the resolvent set of A, and $\sigma_{\mathrm{p}}(A)$ is the set of eigenvalues of A.

[3] A self-adjoint unitary operator γ on a Hilbert space with $\gamma \neq \pm I$ is called a **grading operator** on $\mathscr{H}$.

[4] In general, for a self-adjoint operator A on a Hilbert space $\mathscr{K}$ and a Borel measurable function f on $\mathbb{R}$, the operator $f(A)$ is defined as the operator satisfying the following: $\mathrm{Dom}(f(A)) = \{\Psi \in \mathscr{K} | \int_{\mathbb{R}} |f(\lambda)|^2 d\|E_A(\lambda)\Psi\|^2 < \infty\}$ and $\langle \Phi, f(A)\Psi \rangle = \int_{\mathbb{R}} f(\lambda) d \langle \Phi, E_A(\lambda)\Psi \rangle$, $\Phi \in \mathscr{K}, \Psi \in \mathrm{Dom}(f(A))$. The operator $f(A)$ is symbolically denoted as $f(A) = \int f(\lambda) dE_A(\lambda)$.

(ii) *Suppose that $\sigma_p(Q) \neq \emptyset$. Then, for each $\lambda \in \sigma_p(Q)$, $-\lambda$ is in $\sigma_p(Q)$ and the dimension of $\ker(Q - \lambda)$,[5] the eigenspace of Q with eigenvalue λ, coincides with that of $\ker(Q + \lambda)$, the eigenspace of Q with eigenvalue $-\lambda$:*

$$\dim \ker(Q - \lambda) = \dim \ker(Q + \lambda). \tag{1.4}$$

Proof (i) Condition (i) in Definition 1.1 implies that $\Gamma^{-1} = \Gamma$. This property and condition (iii) in Definition 1.1 imply the operator equality[6]

$$\Gamma Q \Gamma^{-1} = -Q. \tag{1.5}$$

Hence Q and $-Q$ are unitarily equivalent. It follows from the unitary invariance of the spectrum of a linear operator that $\sigma(Q) = \sigma(-Q)$. This means that $\sigma(Q)$ is symmetric with respect to the origin of $\mathbb{R}$.

(ii) By (1.5), for all $z \in \mathbb{C}$, we have $\Gamma(Q - z)\Gamma^{-1} = -(Q + z)$. In particular, for each $\lambda \in \sigma_p(Q)$, this equation implies that $-\lambda \in \sigma_p(Q)$ and $\Gamma \ker(Q - \lambda) = \ker(Q + \lambda)$. Hence (1.4) follows. $\qquad\square$

Since we have

$$H = Q^2, \tag{1.6}$$

it follows that

$$\ker H = \ker Q. \tag{1.7}$$

The spectrum of Q is related to that of H in the way stated in the following theorem:

Theorem 1.2 (i) $\sigma(Q) = \{\pm\sqrt{\mu}|\mu \in \sigma(H)\}$; (ii) $\sigma_p(Q) = \{\pm\sqrt{\mu}|\mu \in \sigma_p(H)\}$.

Proof (i) By (1.3) and the spectral mapping theorem, we have $\sigma(|Q|) = \{\sqrt{\mu}|\mu \in \sigma(H)\}$. We denote by E_Q the spectral measure of the self-adjoint operator Q. Let $\mathscr{E}_+ := \mathrm{Ran}(E_Q((0, \infty)))$ and $\mathscr{E}_- := \mathrm{Ran}(E_Q((-\infty, 0)))$. Then we have the orthogonal decomposition $\mathscr{H} = \mathscr{E}_+ \oplus \ker Q \oplus \mathscr{E}_-$. The operator Q is reduced by $\mathscr{E}_\pm$[7] and $\ker Q$. We denote the reduced part of Q to $\mathscr{E}_\pm$ (resp. $\ker Q$) by

---

[5] For a linear operator A on a Hilbert space, $\ker A := \{\Psi \in \mathrm{Dom}(A)|A\Psi = 0\}$, the kernel of A.

[6] Let A and B be linear operators from a Hilbert space $\mathscr{H}_1$ to a Hilbert space $\mathscr{H}_2$. (i) B is said to be an **extension** of A if $\mathrm{Dom}(A) \subset \mathrm{Dom}(B)$ and $A\Psi = B\Psi$, $\Psi \in \mathrm{Dom}(A)$. In this case, we write $A \subset B$. (ii) A is said to be **equal** to B if $\mathrm{Dom}(A) = \mathrm{Dom}(B)$ and $A\Psi = B\Psi$, $\Psi \in \mathrm{Dom}(A)(= \mathrm{Dom}(B))$. In this case, we write $A = B$. This type of equality is called operator equality. It follows that $A = B$ if and only if $A \subset B$ and $B \subset A$. The notion of extension of linear operator is important in treating linear operators *not everywhere defined*, in particular, *unbounded* linear operators.

[7] A linear operator A on a Hilbert space $\mathscr{H}$ is said to be reduced by a closed subspace $\mathscr{M}$ of $\mathscr{H}$ if, for all $\Psi \in \mathrm{Dom}(A)$, $P_{\mathscr{M}}\Psi$ ($P_{\mathscr{M}}$ is the orthogonal projection to $\mathscr{M}$) is in $\mathrm{Dom}(A)$ and $AP_{\mathscr{M}}\Psi = P_{\mathscr{M}}A\Psi$. In this case, one can define an operator $A_{\mathscr{M}}$ on $\mathscr{M}$ as follows: $\mathrm{Dom}(A_{\mathscr{M}}) := \mathrm{Dom}(A) \cap \mathscr{M}$, $A_{\mathscr{M}}\Psi := A\Psi$, $\Psi \in \mathrm{Dom}(A_{\mathscr{M}})$. The operator $A_{\mathscr{M}}$ is called the reduced part of A to $\mathscr{M}$.

$Q^{(\pm)}$ (resp. $Q^{(0)}$). By functional calculus, we have $Q = Q^{(+)} \oplus Q^{(0)} \oplus Q^{(-)}$ and $|Q| = Q^{(+)} \oplus Q^{(0)} \oplus (-Q^{(-)})$. Hence

$$\sigma(Q) = \sigma(Q^{(+)}) \cup \sigma(Q^{(0)}) \cup \sigma(Q^{(-)}), \sigma(|Q|) = \sigma(Q^{(+)}) \cup \sigma(Q^{(0)}) \cup \sigma(-Q^{(-)}).$$

By Theorem 1.1 (i), $\sigma(\pm Q^{(-)}) \setminus \{0\} = \sigma(\mp Q^{(+)}) \setminus \{0\}$. Therefore $\sigma(|Q|) \setminus \{0\} = \sigma(Q^{(+)}) \setminus \{0\}$. Thus $\sigma(Q) \setminus \{0\} = (\sigma(|Q|) \setminus \{0\}) \cup (\sigma(-|Q|) \setminus \{0\})$. This implies the following: (a) if ker $Q = \{0\}$, then $\sigma(Q) = \sigma(|Q|) \cup \sigma(-|Q|)$. Hence the desired result holds; (b) if ker $Q \neq \{0\}$, then $\sigma(Q^{(0)}) = \{0\}$ and hence, by (1.7), $0 \in \sigma(H)$. Therefore $0 \in \sigma(Q)$ and the desired result holds.

(ii) (1.6) implies that $\sigma_{\mathrm{p}}(Q) \subset \{\lambda \in \mathbb{R} | \lambda^2 \in \sigma_{\mathrm{p}}(H)\}$. Conversely, let $\lambda \in \mathbb{R}$ be such that $\lambda^2 \in \sigma_{\mathrm{p}}(H)$. Then there exists a non-zero vector $\Psi \in \mathrm{Dom}(H)$ such that $H\Psi = \lambda^2 \Psi$. Hence, by (1.6), $(Q - \lambda)(Q + \lambda)\Psi = 0$. This implies that $-\lambda \in \sigma_{\mathrm{p}}(Q)$ or $\lambda \in \sigma_{\mathrm{p}}(Q)$. Hence, by Theorem 1.1 (ii), $\lambda \in \sigma_{\mathrm{p}}(Q)$. Therefore $\sigma(Q) = \{\lambda \in \mathbb{R} | \lambda^2 \in \sigma_{\mathrm{p}}(H)\}$. Thus the desired result holds. $\qquad\square$

1.3   Orthogonal Decomposition of State Vectors

By (1.2), $\mathscr{H}$ has the orthogonal decomposition

$$\mathscr{H} = \mathscr{H}_+ \oplus \mathscr{H}_- = \{(\Psi_+, \Psi_-) | \Psi_\pm \in \mathscr{H}_\pm\} \tag{1.8}$$

with $\mathscr{H}_+ := \ker(\Gamma - 1)$, $\mathscr{H}_- := \ker(\Gamma + 1)$. The closed subspaces $\mathscr{H}_+$ and $\mathscr{H}_-$ are called the **bosonic subspace** and the **fermionic subspace** respectively. A non-zero vector in $\mathscr{H}_+$ (resp. $\mathscr{H}_-$) is called a **bosonic** (resp. **fermionic**) **state**. Since $\Gamma\Psi_\pm = \pm\Psi_\pm$ for all $\Psi_\pm \in \mathscr{H}_\pm$, we call the operator Γ the **state–sign operator**.[8]

Let $P_\pm$ be the orthogonal projections onto $\mathscr{H}_\pm$. Then we have

$$\Gamma = P_+ - P_-. \tag{1.9}$$

By this equation and the relation $P_+ + P_- = I$, we obtain

$$P_\pm = \frac{1}{2}(I \pm \Gamma). \tag{1.10}$$

---

[8] In the physics literature, Γ is often written as "$(-1)^F$".

1.4 Operator Matrix Representations

Since $\mathscr{H}$ has the orthogonal decomposition (1.8), a linear operator on $\mathscr{H}$ may have an operator matrix representation with respect to (1.8) (see Appendix A). By (1.9) and Remark A.1 in Appendix A, we have

$$\Gamma = I \oplus (-I) = \begin{pmatrix} I & 0 \\ 0 & -I \end{pmatrix}. \tag{1.11}$$

We next derive the operator matrix representation of Q with respect to (1.8).

Lemma 1.1 *For all $\Psi \in \mathrm{Dom}(Q)$, $P_{\pm}\Psi \in \mathrm{Dom}(Q)$ and $QP_{+}\Psi = P_{-}Q\Psi$, $QP_{-}\Psi = P_{+}Q\Psi$. In particular, Q maps $\mathrm{Dom}(Q) \cap \mathscr{H}_{\pm}$ to $\mathscr{H}_{\mp}$.*

Proof An easy exercise (use (1.10) and (1.1)). $\square$

For two Hilbert spaces $\mathscr{H}$ and $\mathscr{K}$, we denote by $\mathfrak{C}(\mathscr{H}, \mathscr{K})$ the set of densely defined closed linear operators from $\mathscr{H}$ to $\mathscr{K}$.

It follows from Lemma 1.1 and Lemma A.1 (i)–(ii) that the operator matrix representation of Q with respect to (1.8) takes the form:

$$Q = \begin{pmatrix} 0 & Q_- \\ Q_+ & 0 \end{pmatrix}$$

with $Q_+ \in \mathfrak{C}(\mathscr{H}_+, \mathscr{H}_-)$ and $Q_- \in \mathfrak{C}(\mathscr{H}_-, \mathscr{H}_+)$. The self-adjointness of Q and Theorem A.1 (ii) imply that $Q_- = Q_+^*$, the adjoint of Q_+. Hence we obtain

$$Q = \begin{pmatrix} 0 & Q_+^* \\ Q_+ & 0 \end{pmatrix}. \tag{1.12}$$

Applying Lemma A.2 to the case $A = B = Q$ and using (1.6) and (1.12), we obtain the operator matrix representation of H:

$$H = \begin{pmatrix} H_+ & 0 \\ 0 & H_- \end{pmatrix} = H_+ \oplus H_- \tag{1.13}$$

with

$$H_+ := Q_+^* Q_+, \quad H_- := Q_+ Q_+^*. \tag{1.14}$$

Hence H is reduced by $\mathscr{H}_{\pm}$ and the reduced part of H to $\mathscr{H}_+$ (resp. $\mathscr{H}_-$) is given by H_+ (resp. H_-). The operators $H_{\pm}$ are non-negative self-adjoint operators. We call H_+ (resp. H_-) the **bosonic** (resp. **fermionic**) **Hamiltonian**. It follows from (1.14) that $H_{\pm} = |Q_{\pm}|^2$ and hence

$$H_{\pm}^{1/2} = |Q_{\pm}|. \tag{1.15}$$

1.5 Construction of SQM

In this section, we describe two methods to construct an SQM in an abstract frame-work.

1.5.1 Method I

Let $(\mathscr{K}_+, \mathscr{K}_-, A)$ be a triple consisting of Hilbert spaces $\mathscr{K}_\pm$ and $A \in \mathfrak{C}(\mathscr{K}_+, \mathscr{K}_-)$. Then the triple yields the direct sum Hilbert space $\mathscr{K} := \mathscr{K}_+ \oplus \mathscr{K}_-$ and operators

$$h_A := A^*A \oplus AA^* = \begin{pmatrix} A^*A & 0 \\ 0 & AA^* \end{pmatrix}, \quad q_A := \begin{pmatrix} 0 & A^* \\ A & 0 \end{pmatrix}.$$

It is easy to see that h_A and q_A are self-adjoint [22, Proposition B.1, Theorem B.2]. It follows that $h_A = q_A^2$. It is obvious that the operator $\Gamma_{\mathscr{K}} := I \oplus (-I)$ on $\mathscr{K}$ is a grading operator. Thus $(\mathscr{K}, \Gamma_{\mathscr{K}}, q_A, h_A)$ is an $N = 1$ SQM. This method can easily be extended to construct an SQM with supersymmetry of degree $N \geq 2$.

1.5.2 Method II

In applications to concrete models of SQM, there may be the case where only the symmetricity[9] of the operator which is expected to be a supercharge is known and it may be non-trivial to prove or disprove its (essential) self-adjointness. In this case, one may proceed as follows. Let $(\mathscr{H}, \Gamma, Q)$ be a triple consisting of a Hilbert space $\mathscr{H}$, a grading operator Γ on $\mathscr{H}$ and a closed symmetric operator Q such that Q anti-commutes with Γ on $\mathrm{Dom}(Q)$: for all $\Psi \in \mathrm{Dom}(Q)$, $\Gamma\Psi$ is in $\mathrm{Dom}(Q)$ and

$$\Gamma Q\Psi + Q\Gamma\Psi = 0. \tag{1.16}$$

Such an operator Q is called an **abstract Dirac operator** with respect to Γ. We have the orthogonal decomposition (1.8). Then Q has the operator matrix representation

$$Q = \begin{pmatrix} 0 & Q_- \\ Q_+ & 0 \end{pmatrix},$$

---

[9] A linear operator S on a Hilbert space is said to be symmetric if $\mathrm{Dom}(S)$ is dense and, for all $\Psi, \Phi \in \mathrm{Dom}(S)$, $\langle \Phi, S\Psi \rangle = \langle S\Phi, \Psi \rangle$. It follows that S is symmetric if and only if $\mathrm{Dom}(S)$ is dense and $S \subset S^*$ (i.e., S^* is an extension of S).

where $Q_\pm := Q \upharpoonright \mathrm{Dom}(Q) \cap \mathcal{H}_\pm$ (the restriction of Q to $\mathrm{Dom}(Q) \cap \mathcal{H}_\pm$). It follows that $Q_\pm$ are densely defined closed linear operators. Hence, by Theorem A.2 in Appendix A, the operators

$$Q_1 := \begin{pmatrix} 0 & Q_+^* \\ Q_+ & 0 \end{pmatrix}, \quad Q_2 := \begin{pmatrix} 0 & Q_- \\ Q_-^* & 0 \end{pmatrix}$$

are self-adjoint extensions of Q. Moreover, using (1.11) and Lemma A.2 in Appendix A, one can show that (1.16) holds with Q replaced by Q_1 and Q_2 respectively. Thus we have two SQM $(\mathcal{H}, \Gamma, Q_1, H_1)$ and $(\mathcal{H}, \Gamma, Q_2, H_2)$ with $H_1 := Q_1^2$ and $H_2 := Q_2^2$. If Q is self-adjoint, then $Q = Q_1 = Q_2$. But, if Q is not self-adjoint, then $Q_1 \neq Q_2$.

1.6 Spectral Supersymmetry

Let $(\mathcal{H}, \Gamma, \{Q_i\}_{i=1}^N, H)$ be an SQM. Then, by (1.13), we have

$$\sigma(H) = \sigma(H_+) \cup \sigma(H_-) \subset [0, \infty), \quad \sigma_\mathrm{p}(H) = \sigma_\mathrm{p}(H_+) \cup \sigma_\mathrm{p}(H_-).$$

Moreover, there exist characteristic structures between the spectra of the bosonic Hamiltonian H_+ and those of the fermionic Hamiltonian H_-:

Theorem 1.3 (spectral supersymmetry)

$$\sigma(H) \setminus \{0\} = \sigma(H_+) \setminus \{0\} = \sigma(H_-) \setminus \{0\},$$
$$\sigma_\mathrm{p}(H) \setminus \{0\} = \sigma_\mathrm{p}(H_+) \setminus \{0\} = \sigma_\mathrm{p}(H_-) \setminus \{0\}.$$

Moreover, for each $E \in \sigma_\mathrm{p}(H_+) \setminus \{0\}$, define $U_E : \ker(H_+ - E) \to \mathcal{H}_-$ by

$$U_E \Psi := \frac{1}{\sqrt{E}} Q_+ \Psi, \quad \Psi \in \ker(H_+ - E).$$

Then $\mathrm{Ran}(U_E) = \ker(H_- - E)$ and U_E is a unitary operator from $\ker(H_+ - E)$ to $\ker(H_- - E)$. In particular,

$$\dim \ker(H_+ - E) = \dim \ker(H_- - E)$$

and each positive eigenvalue E of H is degenerate with $\dim \ker(H - E)$ being even or ∞.

Theorem 1.3 can be proved by an application of a general theorem:

Theorem 1.4 (Deift's theorem) *Let $\mathscr{H}$, $\mathscr{K}$ be Hilbert spaces and $A \in \mathfrak{C}(\mathscr{H}, \mathscr{K})$. Then A^*A and AA^* are non-negative self-adjoint operators on $\mathscr{H}$ and $\mathscr{K}$ respectively and*

$$\sigma(A^*A) \setminus \{0\} = \sigma(AA^*) \setminus \{0\}, \tag{1.17}$$
$$\sigma_p(A^*A) \setminus \{0\} = \sigma_p(AA^*) \setminus \{0\}.$$

Moreover, for each $\lambda \in \sigma_p(A^*A) \setminus \{0\}$, the operator $U_\lambda : \ker(A^*A - \lambda) \to \mathscr{K}$ defined by

$$U_\lambda \Psi := \frac{1}{\sqrt{\lambda}} A\Psi, \quad \Psi \in \ker(A^*A - \lambda)$$

is a unitary operator from $\ker(A^*A - \lambda)$ to $\ker(AA^* - \lambda)$. In particular,

$$\dim \ker(A^*A - \lambda) = \dim \ker(AA^* - \lambda).$$

For proof of this theorem, we refer the reader to [31] or [22, Theorem 7.23], [18, Theorem 9.8].

Proof of Theorem 1.3

We have (1.13) and (1.14). Hence we need only to apply Theorem 1.4 to the case where $\mathscr{H} = \mathscr{H}_+$, $\mathscr{K} = \mathscr{H}_-$ and $A = Q_+$. □

The degenerate structure of each positive eigenvalue of the supersymmetric Hamiltonian H described in Theorem 1.3 is interesting to note.

1.7   Ground States

In general, for a self-adjoint operator A on a Hilbert space, which is bounded from below, the infimum of the spectrum of A

$$E_0(A) := \inf \sigma(A) > -\infty$$

is called the **lowest energy** of A.[10] If $E_0(A) > 0$, then A is said to be **strictly positive**.

If $E_0(A)$ is an eigenvalue of A (i.e., $E_0(A) \in \sigma_p(A)$), then each non-zero vector in $\ker(A - E_0(A))$ (resp. $E_0(A)$) is called a **ground state** (resp. the **ground state**

---

[10] Originally this term is used only for the case where A denotes a quantum mechanical Hamiltonian.

energy) of A. In this case, we say that A has a ground state. If $\dim \ker(A - E_0(A)) = 1$ (resp. $\dim \ker(A - E_0(A)) \geq 2$), then the ground sate of A is said to be **unique** (resp. **degenerate**). If $E_0(A) = 0 \in \sigma_p(A)$, then a ground state of A is called a **zero-energy ground state**.

Theorem 1.3 yields the following result:

Theorem 1.5 *If H is strictly positive and has a ground state, then the ground state of H is degenerate with $\dim \ker(H - E_0(H))$ being an even number or ∞.*

With regard to zero-energy ground states of H, we have by (1.13)

$$\ker H = \ker H_+ \oplus \ker H_-. \tag{1.18}$$

Hence, if $\ker H \neq \{0\}$ (i.e., H has a zero-energy ground state), then $\ker H_+ \neq \{0\}$ or $\ker H_- \neq \{0\}$.

1.8 Spontaneous Supersymmetry Breaking and an Index Formula

Physically, the supercharge Q is interpreted as the generator of supersymmetry in the SQM under consideration. Hence a state vector eliminated by Q is regarded as a state with supersymmetry. Based on this picture, a non-zero vector $\Psi \in \mathrm{Dom}(Q)$ such that $Q\Psi = 0$ (if it exists) is called a **supersymmetric state**. Hence the set of supersymmetric states is given by $\ker Q \setminus \{0\}$.

If $\ker Q = \{0\}$, then there exist no supersymmetric states. In this case, we say that the supersymmetry is **spontaneously broken**.

Using (1.7), one can characterize the spontaneous supersymmetry breaking in terms of the supersymmetric Hamiltonian H:

Theorem 1.6 *The supersymmetry is spontaneously broken if and only if H has no zero-energy ground states. In particular, if H is strictly positive, then the supersymmetry is spontaneously broken.*

It follows from (1.12) that

$$\ker Q = \ker Q_+ \oplus \ker Q_+^*. \tag{1.19}$$

Hence

$$\dim \ker Q = \dim \ker Q_+ + \dim \ker Q_+^* \in \mathbb{Z}_+ \cup \{\infty\},$$

where $\mathbb{Z}_+ := \{0\} \cup \mathbb{N}$ (the set of non-negative integers).

In general, for a densely defined linear operator A from a Hilbert space to a Hilbert space such that at least one of $\ker A$ and $\ker A^*$ is finite-dimensional, one can define an object

$$\text{ind } A := \dim \ker A - \dim \ker A^* \in \mathbb{Z} \cup \{\pm\infty\},$$

where $\mathbb{Z}$ is the set of integers. This object is called the **index** of A. It is well known that, even if $\dim \ker A$ and $\dim \ker A^*$ are not calculated, there exist cases where ind A can be calculated; the so-called "index theorems" give methods of such calculations (see, e.g., [33]).

A necessary condition for the supersymmetry to be spontaneously broken is given as follows:

Proposition 1.1 *If the supersymmetry is spontaneously broken, then* ind $Q_+ = 0$.

Proof The present assumption is equivalent to $\ker Q = \{0\}$. Hence, by (1.19), $\ker Q_+ = \{0\}$ and $\ker Q_+^* = \{0\}$. Therefore $\dim \ker Q_+ = 0$ and $\dim \ker Q_+^* = 0$. Thus ind $Q_+ = 0$. □

It follows from (1.14) that

$$\ker H_+ = \ker Q_+, \quad \ker H_- = \ker Q_+^*, \tag{1.20}$$

which, together with (1.18), imply that

$$\ker H = \ker Q_+ \oplus \ker Q_+^*.$$

The difference between the number of bosonic zero-energy states and that of fermionic zero-energy states

$$\Delta_W := \dim \ker H_+ - \dim \ker H_-$$

is called the **Witten index**, provided that one of $\dim \ker H_+$ and $\dim \ker H_-$ is finite. By (1.20), we have

$$\text{ind } Q_+ = \Delta_W.$$

By this relation and Proposition 1.1, we obtain the following corollary:

Corollary 1.1 *If the supersymmetry is spontaneously broken, then* $\Delta_W = 0$.

In relation to the notion of index of a linear operator, we here recall the definition of a (semi-) Fredholm operator acting in a Hilbert space.

An operator $T \in \mathfrak{C}(\mathcal{H}, \mathcal{K})$ ($\mathcal{H}$ and $\mathcal{K}$ are Hilbert spaces) is said to be **semi-Fredholm** if $\text{Ran}(T)$ is closed and at least one of $\dim \ker T$ and $\dim \ker T^*$ is finite. A semi-Fredholm operator T is called a **Fredholm operator** if both of $\dim \ker T$ and $\dim \ker T^*$ are finite. The importance of (semi-)Fredholm operator lies in that some stability theorems hold [45, Chapter IV, §5].[11]

The index of Q_+ may be computed in terms of the heat semi-group $\{e^{-\beta H}\}_{\beta \geq 0}$ generated by the supersymmetric Hamiltonian H:

---

[11] For a general theory of index of (semi-) Fredholm operator, see, e.g., [33].

Theorem 1.7 *Suppose that, for some constant $\beta_0 > 0$, $e^{-\beta_0 H}$ is trace class. Then, for all $\beta \geq \beta_0$, $e^{-\beta H}$ is trace class and Q_+ is Fredholm with*

$$\text{ind } Q_+ = \text{Tr}\,(\Gamma e^{-\beta H}), \tag{1.21}$$

where Tr *denotes trace. The right-hand side is independent of β.*

Proof The operator $e^{-\beta_0 H}$ is bounded, injective, non-negative and self-adjoint. It follows that $\sigma(e^{-\beta_0 H}) \subset [0, 1]$. Since a trace class operator is compact, $e^{-\beta_0 H}$ is compact. Hence the spectrum of $e^{-\beta_0 H}$ consists of only positive eigenvalues λ_n, $n \geq 1$ ($1 \geq \lambda_1 > \lambda_2 > \cdots > 0$) with a finite multiplicity m_n. By the spectral mapping theorem, the spectrum of H is given by $\{\varepsilon_n\}_n \subset [0, \infty)$ with $\varepsilon_n = -\beta_0^{-1} \log \lambda_n$ and the multiplicity of ε_n being equal to m_n. In particular, $\dim \ker H < \infty$. Hence, by (1.20), $\dim \ker Q_+ < \infty$ and $\dim \ker Q_+^* < \infty$. By (1.15), we have $\|Q_+\Psi\| = \|H_+^{1/2}\Psi\| \geq \sqrt{\delta}\|\Psi\|$, $\Psi \in (\ker Q_+)^\perp \cap \text{Dom}(Q_+)$, where $\delta := \min \sigma(H_+) \setminus \{0\} > 0$. This implies that $\text{Ran}(Q_+)$ is closed. Thus Q_+ is Fredholm. For all $\beta \geq \beta_0$, $\text{Tr}\,e^{-\beta H} = \sum_{n \geq 1} m_n e^{-\beta \varepsilon_n} \leq \sum_{n \geq 1} m_n e^{-\beta_0 \varepsilon_n} = \text{Tr}\,e^{-\beta_0 H} < \infty$. Hence $e^{-\beta H}$ is trace class for all $\beta \geq \beta_0$. Moreover, using by (1.8) and Theorem 1.3, we have

$$\text{Tr}\,(\Gamma e^{-\beta H}) = \text{Tr}\,e^{-\beta H_+} - \text{Tr}\,e^{-\beta H_-}$$

$$= \dim \ker H_+ - \dim \ker H_- + \sum_{\varepsilon_n > 0} m_n e^{-\beta \varepsilon_n} - \sum_{\varepsilon_n > 0} m_n e^{-\beta \varepsilon_n}$$

$$= \dim \ker H_+ - \dim \ker H_- = \Delta_W = \text{ind } Q_+.$$

Thus (1.21) holds. $\square$

Chapter 2
Elements of the Theory of Fock Spaces

Abstract We review the theory of Fock spaces within the scope of the following chapters (for more details, see [22]).

2.1 Full Fock Space

Let $\mathscr{H}$ be a Hilbert space. For each $n \in \mathbb{N}$, we denote the n-fold tensor product Hilbert space of $\mathscr{H}$ by $\otimes^n \mathscr{H}$. We set $\otimes^0 \mathscr{H} := \mathbb{C}$. The infinite direct sum Hilbert space of $\otimes^n \mathscr{H}$ $(n = 0, 1, 2, \ldots)$

$$\mathscr{F}(\mathscr{H}) := \bigoplus_{n=0}^{\infty} \otimes^n \mathscr{H} = \left\{ \Psi = \{\Psi^{(n)}\}_{n=0}^{\infty} \Big| \Psi^{(n)} \in \otimes^n \mathscr{H}, \, n \geq 0, \, \sum_{n=0}^{\infty} \|\Psi^{(n)}\|^2 < \infty \right\}$$

is called the **full Fock space** over $\mathscr{H}$. The *algebraic* infinite direct sum of $\otimes^n \mathscr{H}$ $(n = 0, 1, 2, \ldots)$

$$\mathscr{F}_0(\mathscr{H}) := \hat{\oplus}_{n=0}^{\infty} \otimes^n \mathscr{H}$$
$$= \left\{ \Psi = \{\Psi^{(n)}\}_{n=0}^{\infty} \Big| \Psi^{(n)} \in \otimes^n \mathscr{H}, \, n \geq 0, \, \exists n_0 \in \mathbb{N} \, (\Psi^{(n)} = 0, \, \forall n \geq n_0) \right\}$$

is a dense subspace of $\mathscr{F}(\mathscr{H})$. The subspace $\mathscr{F}_0(\mathscr{H})$ is called the **finite particle subspace** of $\mathscr{F}(\mathscr{H})$.

For each subspace $\mathscr{D}$ of $\mathscr{H}$, we denote by $\hat{\otimes}^n \mathscr{D}$ the n-fold *algebraic* tensor product of $\mathscr{D}$. It follows that, if $\mathscr{D}$ is dense in $\mathscr{H}$, then $\hat{\otimes}^n \mathscr{D}$ is dense in $\otimes^n \mathscr{H}$. We set $\hat{\otimes}^0 \mathscr{D} := \mathbb{C}$. If $\mathscr{D}$ is dense in $\mathscr{H}$, then the *algebraic* infinite direct sum of $\hat{\otimes}^n \mathscr{D}$ $(n = 0, 1, 2, \ldots)$

$$\mathscr{F}_{\text{fin}}(\mathscr{D}) := \hat{\bigoplus}_{n=0}^{\infty} \hat{\otimes}^n \mathscr{D}$$

is dense in $\mathscr{F}(\mathscr{H})$. It is obvious that $\mathscr{F}_{\text{fin}}(\mathscr{D}) \subset \mathscr{F}_0(\mathscr{H})$.

A. Arai, *Infinite-Dimensional Dirac Operators and Supersymmetric Quantum Fields*, SpringerBriefs in Mathematical Physics, https://doi.org/10.1007/978-981-19-5678-2_2

The vector $\Omega_{\mathcal{H}} \in \mathscr{F}(\mathcal{H})$ defined by

$$\Omega_{\mathcal{H}}^{(0)} := 1, \quad \Omega^{(n)} = 0, \quad n \geq 1$$

is called the **Fock vacuum** in $\mathscr{F}(\mathcal{H})$. This vector plays an important role in the theory of Fock spaces.

2.2 Boson Fock Space

For each $n \in \mathbb{N}$, we denote by S_n the symmetrization operator (symmetrizer) on $\otimes^n \mathcal{H}$, i.e., S_n is the bounded linear operator on $\otimes^n \mathcal{H}$ such that

$$S_n(\psi_1 \otimes \cdots \otimes \psi_n) = \frac{1}{n!} \sum_{\sigma \in \mathfrak{S}_n} \psi_{\sigma(1)} \otimes \cdots \otimes \psi_{\sigma(n)}, \quad \psi_i \in \mathcal{H}, \ i = 1, \ldots, n,$$

where $\mathfrak{S}_n := \{\sigma : \{1, \ldots, n\} \to \{1, \ldots, n\} | \sigma \text{ is injective}\}$ denotes the symmetry group of order n (the permutation group of order n). It is shown that S_n is an orthogonal projection [22, Theorem 2.9(ii)]. Hence its range

$$\otimes_s^n \mathcal{H} := \text{Ran}(S_n)$$

is a closed subspace of $\otimes^n \mathcal{H}$. This closed subspace is called the n-**fold symmetric tensor product Hilbert space** of $\mathcal{H}$. We set $\otimes_s^0 \mathcal{H} := \mathbb{C}$. In the context of quantum field theory, $\otimes_s^n \mathcal{H}$ gives an abstract form of Hilbert spaces of state vectors of n identical bosons.

The infinite direct sum Hilbert space of $\otimes_s^n \mathcal{H}$ $(n = 0, 1, 2, \ldots)$

$$\mathscr{F}_b(\mathcal{H}) := \bigoplus_{n=0}^{\infty} \otimes_s^n \mathcal{H},$$

which is a closed subspace of $\mathscr{F}(\mathcal{H})$, is called the **boson** (or **symmetric**) **Fock space** over $\mathcal{H}$. The vector space

$$\mathscr{F}_{b,0}(\mathcal{H}) := \mathscr{F}_b(\mathcal{H}) \cap \mathscr{F}_0(\mathcal{H}),$$

as a subspace of $\mathscr{F}_b(\mathcal{H})$, is called the **bosonic finite particle subspace** of $\mathscr{F}_b(\mathcal{H})$.

For each subspace $\mathscr{D}$ of $\mathcal{H}$, the vector space

$$\hat{\otimes}_s^n \mathscr{D} := S_n(\hat{\otimes}^n \mathscr{D})$$

is called the n-fold *algebraic* symmetric tensor product of $\mathscr{D}$. If $\mathscr{D}$ is dense in $\mathscr{H}$, then $\hat{\otimes}_s^n \mathscr{D}$ is dense in $\otimes_s^n \mathscr{H}$. We set $\otimes_s^0 \mathscr{D} := \mathbb{C}$. We denote by $\mathscr{F}_{b,fin}(\mathscr{D})$ the *algebraic* infinite direct sum of $\hat{\otimes}_s^n \mathscr{D}$ ($n = 0, 1, 2, \ldots$):

$$\mathscr{F}_{b,fin}(\mathscr{D}) := \hat{\bigoplus}_{n=0}^{\infty} \hat{\otimes}_s^n \mathscr{D}.$$

It follows that, if $\mathscr{D}$ is dense in $\mathscr{H}$, then $\mathscr{F}_{b,fin}(\mathscr{D})$ is dense in $\mathscr{F}_b(\mathscr{H})$.

2.3 Fermion Fock Space

For each $n \in \mathbb{N}$, we denote by A_n the anti-symmetrization operator (anti-symmetrizer) on $\otimes^n \mathscr{H}$, i.e., A_n is the bounded linear operator on $\otimes^n \mathscr{H}$ such that

$$A_n(\psi_1 \otimes \cdots \otimes \psi_n) = \frac{1}{n!} \sum_{\sigma \in \mathfrak{S}_n} \text{sgn}(\sigma) \psi_{\sigma(1)} \otimes \cdots \otimes \psi_{\sigma(n)}, \quad \psi_i \in \mathscr{H}, \ i = 1, \ldots, n,$$

where $\text{sgn}(\sigma)$ is the sign of the permutation σ. It is shown that A_n is an orthogonal projection [22, Theorem 2.9(ii)]. Hence its range

$$\wedge^n(\mathscr{H}) := \text{Ran}(A_n)$$

is a closed subspace of $\otimes^n \mathscr{H}$. This closed subspace is called the n-**fold anti-symmetric tensor product Hilbert space** of $\mathscr{H}$. Each element of $\wedge^n(\mathscr{H})$ is called an anti-symmetric tensor of order n. We set $\wedge^0(\mathscr{H}) := \mathbb{C}$. In the context of quantum field theory, $\wedge^n(\mathscr{H})$ gives an abstract form of Hilbert spaces of state vectors of n identical fermions.

The infinite direct sum Hilbert space of $\wedge^n(\mathscr{H})$ ($n = 0, 1, 2, \ldots$)

$$\mathscr{F}_f(\mathscr{H}) := \bigoplus_{n=0}^{\infty} \wedge^n(\mathscr{H}), \tag{2.1}$$

which is a closed subspace of $\mathscr{F}(\mathscr{H})$, is called the **fermion** (or **anti-symmetric**) **Fock space** over $\mathscr{H}$. The vector space

$$\mathscr{F}_{f,0}(\mathscr{H}) := \mathscr{F}_f(\mathscr{H}) \cap \mathscr{F}_0(\mathscr{H}),$$

as a subspace of $\mathscr{F}_f(\mathscr{H})$, is called the **fermionic finite particle subspace** of $\mathscr{F}_f(\mathscr{H})$.

For each subspace $\mathscr{D}$ of $\mathscr{H}$, the vector space

$$\hat{\wedge}^n(\mathscr{D}) := A_n(\hat{\otimes}^n \mathscr{D})$$

is called the n-fold *algebraic* anti-symmetric tensor product of $\mathscr{D}$. If $\mathscr{D}$ is dense in $\mathscr{H}$, then $\hat{\wedge}^n(\mathscr{D})$ is dense in $\wedge^n(\mathscr{H})$. We set $\wedge^0(\mathscr{D}) := \mathbb{C}$. We denote by $\mathscr{F}_{f,fin}(\mathscr{D})$ the *algebraic* infinite direct sum of $\hat{\wedge}^n(\mathscr{D})$ ($n = 0, 1, 2, \ldots$):

$$\mathscr{F}_{f,fin}(\mathscr{D}) := \hat{\bigoplus}_{n=0}^{\infty} \hat{\wedge}^n(\mathscr{D}).$$

It follows that, if $\mathscr{D}$ is dense in $\mathscr{H}$, then $\mathscr{F}_{f,fin}(\mathscr{D})$ is dense in $\mathscr{F}_f(\mathscr{H})$.

We have the orthogonal decomposition

$$\mathscr{F}_f(\mathscr{H}) = \mathscr{F}_{f,+}(\mathscr{H}) \oplus \mathscr{F}_{f,-}(\mathscr{H}) \tag{2.2}$$

with $\mathscr{F}_{f,+}(\mathscr{H}) := \oplus_{p=0}^{\infty} \wedge^{2p}(\mathscr{H})$, $\mathscr{F}_{f,-}(\mathscr{H}) := \oplus_{p=0}^{\infty} \wedge^{2p+1}(\mathscr{H})$.

2.4 Second Quantization Operators on the Full Fock Space

Let A_n be a linear operator on a Hilbert space $\mathscr{H}_n$ ($n \in \mathbb{Z}_+$). Then the infinite direct sum $\oplus_{n=0}^{\infty} A_n$ of $\{A_n\}_n$ on the Hilbert space $\oplus_{n=0}^{\infty} \mathscr{H}_n$ is defined as follows:

$$\mathrm{Dom}(\oplus_{n=0}^{\infty} A_n) := \left\{ \Psi \in \oplus_{n=0}^{\infty} \mathscr{H}_n \mid \Psi^{(n)} \in \mathrm{Dom}(A_n), \, n \geq 0, \, \sum_{n=0}^{\infty} \|A_n \Psi^{(n)}\|^2 < \infty \right\},$$

$$(\oplus_{n=0}^{\infty} A_n \Psi)^{(m)} := A_m \Psi^{(m)}, \quad \Psi \in \mathrm{Dom}(\oplus_{n=0}^{\infty} A_n), \, m \geq 0.$$

Let T be a densely defined closed linear operator on $\mathscr{H}$ and I be the identity operator on $\mathscr{H}$. Then, for each $j = 1, \ldots, n$, one has a tensor product operator [22, §3.7]:

$$T_j := I \otimes \cdots \otimes I \otimes \overset{j\text{th}}{T} \otimes I \otimes \cdots \otimes I$$

acting in $\otimes^n \mathscr{H}$, which is densely defined closed. The set $\{T_j\}_{j=1}^n$ yields a densely defined closed linear operator on $\otimes^n \mathscr{H}$:

$$T^{(n)} := \overline{\sum_{j=1}^{n} T_j}, \tag{2.3}$$

where, for a closable linear operator L, $\bar{L}$ denotes the closure of L. We set $T^{(0)} := 0$ acting in $\otimes^0 \mathscr{H}$. Then the infinite direct sum of $\{T^{(n)}\}_n$

$$d\Gamma(T) := \oplus_{n=0}^{\infty} T^{(n)}$$

is a densely defined closed linear operator on $\mathscr{F}(\mathscr{H})$. The operator $d\Gamma(T)$ is called the **second quantization operator** (or simply second quantization) of T on the full Fock space $\mathscr{F}(\mathscr{H})$. It is known that, if T be self-adjoint, then $d\Gamma(T)$ is self-adjoint [22, Theorem 4.8]. It is obvious that $\Omega_{\mathscr{H}} \in \text{Dom}(d\Gamma(T))$ and

$$d\Gamma(T)\Omega_{\mathscr{H}} = 0. \tag{2.4}$$

For a general linear operator A on $\mathscr{H}$ (not necessarily a densely defined closed operator), one has the operator

$$A_{\text{alg}}^{(n)} := \sum_{j=1}^{n} I \hat{\otimes} \cdots \hat{\otimes} I \hat{\otimes} \overset{j\text{th}}{A} \hat{\otimes} I \hat{\otimes} \cdots \hat{\otimes} I \quad (\hat{\otimes} \text{ means algebraic tensor product})$$

with $\text{Dom}(A_{\text{alg}}^{(n)}) := \hat{\otimes}^{n}\text{Dom}(A)$. Then one can define the algebraic infinite direct sum of $\{A_{\text{alg}}^{(n)}\}_{n=0}^{\infty}$ ($A_{\text{alg}}^{(0)} := 0$)

$$d\Gamma^{(\text{alg})}(A) := \hat{\oplus}_{n=0}^{\infty} A_{\text{alg}}^{(n)}. \tag{2.5}$$

We call the operator $d\Gamma^{(\text{alg})}(A)$ the **algebraic second quantization** of A.

There is another type of second quantization operator on the full Fock space $\mathscr{F}(\mathscr{H})$. It is defined by the infinite direct sum of n-fold tenor product operators $\otimes^{n}T := T \otimes T \otimes \cdots \otimes T$ ($n = 0, 1, 2, \ldots$) of T with $\otimes^{0}T := 1$:

$$\Gamma(T) := \oplus_{n=0}^{\infty} \otimes^{n} T.$$

We call it the Γ-**operator** of T or the **second quantization of second kind** of T [17, §3.3]. If T is unitary, then so is $\Gamma(T)$. A relation between $d\Gamma(\cdot)$ and $\Gamma(\cdot)$ is given as follows: for each self-adjoint operator T on $\mathscr{H}$,

$$\Gamma(e^{itT}) = e^{itd\Gamma(T)}, \quad t \in \mathbb{R}.$$

For other properties of $\Gamma(\cdot)$, see [22, Theorem 4.11].

2.5 Boson Second Quantization Operators

Let T be a densely defined closed linear operator on $\mathscr{H}$ and $T^{(n)}$ be the operator defined by (2.3). Then one can show that, for all $n \in \mathbb{N}$, $S_n T^{(n)} \subset T^{(n)} S_n$ (i.e., $T^{(n)} S_n$ is an extension of $S_n T^{(n)}$). Since S_n is the orthogonal projection onto $\otimes_{s}^{n} \mathscr{H}$, it follows that $T^{(n)}$ is reduced by $\otimes_{s}^{n} \mathscr{H}$. We denote its reduced part to $\otimes_{s}^{n} \mathscr{H}$ by $T_{s}^{(n)}$. Then one can define a closed linear operator on the boson Fock space $\mathscr{F}_{b}(\mathscr{H})$ by

$$d\Gamma_b(T) := \oplus_{n=0}^{\infty} T_s^{(n)}$$

with $T_s^{(0)} := 0$. This operator is called the **boson second quantization operator** of T. It follows that, if T is self-adjoint, then so is $d\Gamma_b(T)$ [22, Theorem 1.38]).

The boson second quantization operator of the identity I on $\mathscr{H}$

$$N_b := d\Gamma_b(I)$$

is called the **boson number operator**. The name comes from the following easily proved formula:

$$N_b \upharpoonright \otimes_s^n \mathscr{H} = n, \quad n \geq 0.$$

It is easy to see that, for each $n \geq 0$, $\otimes^n T$ is reduced by $\otimes_s^n \mathscr{H}$. We denote its reduced part by $(\otimes^n T)_s$ and define

$$\Gamma_b(T) := \oplus_{n=0}^{\infty} (\otimes^n T)_s$$

acting in $\mathscr{F}_b(\mathscr{H})$. We call it the **boson Γ-operator** of T or the **boson second quantization of second kind** of T. It follows that, if T is unitary, then so is $\Gamma_b(T)$.

2.6   Fermion Second Quantization Operators

One can show that, for all $n \in \mathbb{N}$, $A_n T^{(n)} \subset T^{(n)} A_n$. Since A_n is the orthogonal projection onto $\wedge^n(\mathscr{H})$, it follows that $T^{(n)}$ is reduced by $\wedge^n(\mathscr{H})$. We denote its reduced part to $\wedge^n(\mathscr{H})$ by $d\Gamma_f^{(n)}(T)$. We set $d\Gamma_f^{(0)}(T) := 0$. Then one can define a closed linear operator on the fermion Fock space $\mathscr{F}_f(\mathscr{H})$ by

$$d\Gamma_f(T) := \oplus_{n=0}^{\infty} d\Gamma_f^{(n)}(T).$$

This operator is called the **fermion second quantization operator** of T. It is shown that, if T is self-adjoint, then so is $d\Gamma_f(T)$.

The **fermion number operator** is defined by

$$N_f := d\Gamma_f(I). \tag{2.6}$$

As in the case of the boson number operator, we have

$$N_f \upharpoonright \wedge^n(\mathscr{H}) = n, \quad n \geq 0. \tag{2.7}$$

It is easy to see that $\otimes^n T$ is reduced by $\wedge^n(\mathscr{H})$. We denote its reduced part by $\wedge^n(T)$. Then the operator defined by

$$\Gamma_f(T) := \oplus_{n=0}^{\infty} \wedge^n (T) \tag{2.8}$$

acting in $\mathscr{F}_f(\mathscr{H})$ is called the **fermion Γ-operator** of T or the **fermion second quantization of second kind** of T. If T is unitary, then so is $\Gamma_f(T)$.

2.7 Infinite Determinants

Let T be a trace class operator on $\mathscr{H}$. Then it is shown that, for all $n \geq 0$, $\wedge^n(T)$ is trace class [57, p. 323, Lemma 3]. Moreover, the infinite series

$$\det(1 + T) := \sum_{n=0}^{\infty} \text{Tr} \wedge^n (T)$$

is absolutely convergent [57, p.323, Lemma 4]. The number $\det(1 + T)$ is called the **determinant** of $1 + T$. It follows from (2.8) that $\Gamma_f(T)$ is trace class and

$$\det(1 + T) = \text{Tr} \, \Gamma_f(T).$$

We next consider the case where T is Hilbert–Schmidt. In this case, one needs a regularization to define a notion of determinant of $1 + T$. One can show that the operator

$$R_2(T) := (1 + T)e^{-T} - 1$$

is trace class [63, Lemma 9.1]. Hence one can define

$$\det_2(1 + T) := \det(1 + R_2(T)).$$

This is called the **regularized determinant** of the Hilbert–Schmidt operator T.

2.8 Boson Creation and Annihilation Operators

Let $\mathscr{H}$ be a Hilbert space. Then, for each $f \in \mathscr{H}$, there exists a unique closed linear operator $A(f)$ on the boson Fock space $\mathscr{F}_b(\mathscr{H})$ such that its adjoint $A(f)^*$ takes the following form (see, e.g., [22, §5.7] or [23, §6.4]):

$$\text{Dom}(A(f))^*) = \left\{ \Psi \in \mathscr{F}_b(\mathscr{H}) \Big| \sum_{n=1}^{\infty} \|\sqrt{n} S_n(f \otimes \Psi^{(n-1)})\|^2 < \infty \right\},$$

$$(A(f)^*\Psi)^{(0)} = 0, \quad (A(f)^*\Psi)^{(n)} = \sqrt{n} S_n(f \otimes \Psi^{(n-1)}), \quad n \geq 1, \quad \Psi \in \text{Dom}(A(f)^*).$$

The operator $A(f)$ is called the **boson annihilation operator** with test vector f, while $A(f)^*$ is called the **boson creation operator** with test vector f. These operators have the following properties: for all $f \in \mathscr{H}$, $\mathscr{F}_{b,0}(\mathscr{H}) \subset \mathrm{Dom}(A(f)) \cap \mathrm{Dom}(A(f)^*)$ and the operator set $\{A(f), A(f)^* | f \in \mathscr{H}\}$ obeys the canonical commutation relations (CCR) on $\mathscr{F}_{b,0}(\mathscr{H})$: for all $f, g \in \mathscr{H}$,

$$[A(f), A(g)^*] = \langle f, g \rangle, \tag{2.9}$$
$$[A(f), A(g)] = 0, \quad [A(f)^*, A(g)^*] = 0 \quad \text{on } \mathscr{F}_{b,0}(\mathscr{H}). \tag{2.10}$$

The Fock vacuum $\Omega_{\mathscr{H}} \in \mathscr{F}(\mathscr{H})$ belongs to $\mathscr{F}_b(\mathscr{H})$ and satisfies

$$A(f)\Omega_{\mathscr{H}} = 0, \quad f \in \mathscr{H}. \tag{2.11}$$

It is easy to see that, for all $n \in \mathbb{N}$ and $f_1, \dots, f_n \in \mathscr{H}$,

$$(A(f_1)^* \cdots A(f_n)^* \Omega_{\mathscr{H}})^{(n)} = \sqrt{n!} S_n(f_1 \otimes \cdots \otimes f_n),$$
$$(A(f_1)^* \cdots A(f_n)^* \Omega_{\mathscr{H}})^{(m)} = 0, \quad m \neq n.$$

Hence, for each subspace $\mathscr{D}$ of $\mathscr{H}$, we have

$$\mathscr{F}_{b,\mathrm{fin}}(\mathscr{D}) = \mathrm{span} \left\{ \Omega_{\mathscr{H}}, \left(\prod_{j=1}^{n} A(f_j)^* \right) \Omega_{\mathscr{H}} | n \geq 1, f_1, \dots, f_n \in \mathscr{D} \right\}, \tag{2.12}$$

where, for a subset $\mathscr{M}$ of a vector space, span $\mathscr{M}$ denotes the subspace algebraically spanned by $\mathscr{M}$. It is easy to see that, for any densely defined closed linear operator T on $\mathscr{H}$, $\mathscr{F}_{b,\mathrm{fin}}(\mathrm{Dom}(T)) \subset \mathrm{Dom}(d\Gamma_b(T))$ and, for all $n \in \mathbb{N}$ and $f_1, \dots, f_n \in \mathrm{Dom}(T)$,

$$d\Gamma_b(T) A(f_1)^* \cdots A(f_n)^* \Omega_{\mathscr{H}} = \sum_{j=1}^{n} A(f_1)^* \cdots A(Tf_j)^* \cdots A(f_n)^* \Omega_{\mathscr{H}}. \tag{2.13}$$

In what follows, we denote by $A(f)^{\#}$ either $A(f)$ or $A(f)^*$. It is well known that, for all $f \in \mathscr{H}$. $\mathrm{Dom}(N_b^{1/2}) \subset \mathrm{Dom}(A(f)^{\#})$ [22, Corollary 5.9]. With regard to continuity of $A(f)^{\#}$ in $f \in \mathscr{H}$, we have:

Lemma 2.1 *Let $f_n, f \in \mathscr{H}$ and $\lim_{n \to \infty} f_n = f$. Then, for all $\Psi \in \mathrm{Dom}(N_b^{1/2})$, $\lim_{n \to \infty} A(f_n)^{\#}\Psi = A(f)^{\#}\Psi$.*

Proof See [22, Lemma 5.13(iii)]. □

Let T be a self-adjoint operator on $\mathscr{H}$. Then one can prove the following operator equalities [22, Lemma 5.21]:

$$e^{itd\Gamma_b(T)} A(f)^{\#} e^{-itd\Gamma_b(T)} = A(e^{itT} f)^{\#}, \quad t \in \mathbb{R}, f \in \mathscr{H}, \tag{2.14}$$

If $d\Gamma_b(T)$ represents the Hamiltonian of a quantum system, then (2.14) gives the formula for the time-development of $A(f)^\#$.

2.9 Segal Field Operator

For each $f \in \mathscr{H}$, the symmetric operator

$$\Phi_S(f) := \frac{1}{\sqrt{2}}(A(f)^* + A(f))$$

on $\mathscr{F}_b(\mathscr{H})$ is called the **Segal field operator** with test vector f. It is shown that, for each dense subspace $\mathscr{D}$ of $\mathscr{H}$, $\Phi_S(f)$ is essentially self-adjoint on $\mathscr{F}_{b,\mathrm{fin}}(\mathscr{D})$[23, Theorem 5.22]. For notational simplicity, we denote the closure of $\Phi_S(f)$, which is self-adjoint, by the same symbol $\Phi_S(f)$. It follows from (2.9) and (2.10) that, for all $f, g \in \mathscr{H}$,

$$[\Phi_S(f), \Phi_S(g)] = i\,\mathrm{Im}\,\langle f, g\rangle \quad \text{on } \mathscr{F}_{b,0}(\mathscr{H}), \tag{2.15}$$

where, for a complex number $z \in \mathbb{C}$, $\mathrm{Im}\,z$ denotes the imaginary part of z.

Let T be a self-adjoint operator on $\mathscr{H}$. Then, by (2.14), we have

$$e^{itd\Gamma_b(T)}\Phi_S(f)e^{-itd\Gamma_b(T)} = \Phi_S(e^{itT}f), \quad t \in \mathbb{R}, \quad f \in \mathscr{H}. \tag{2.16}$$

2.10 Isomorphisms Among Boson Fock Spaces

Let $\mathscr{H}'$ be a Hilbert space and $T : \mathscr{H} \to \mathscr{H}'$ be a unitary operator. Then $\otimes^n T$ is a unitary operator from $\otimes_s^n \mathscr{H}$ to $\otimes_s^n \mathscr{H}'$. Hence

$$\Gamma_b(T) := \oplus_{n=0}^\infty \otimes^n T$$

with convention $\otimes^0 T := 1$ is a unitary operator from $\mathscr{F}_b(\mathscr{H})$ to $\mathscr{F}_b(\mathscr{H}')$. Therefore $\mathscr{F}_b(\mathscr{H})$ and $\mathscr{F}_b(\mathscr{H}')$ are isomorphic under $\Gamma_b(T)$. It is easy to see that

$$\Gamma_b(T)A_\mathscr{H}(f)\Gamma_b(T)^{-1} = A_{\mathscr{H}'}(Tf), \quad f \in \mathscr{H},$$

where $A_\mathscr{H}(\cdot)$ denotes the annihilation operator on $\mathscr{F}_b(\mathscr{H})$.

2.11 Fermion Creation and Annihilation Operators

Let $\mathcal{K}$ be a Hilbert space. Then, for each $u \in \mathcal{K}$, there exists a unique everywhere defined bounded linear operator $B(u)$ on the fermion Fock space $\mathcal{F}_f(\mathcal{K})$ such that its adjoint $B(u)^*$ takes the following form (see, e.g., [22, §6.6 and §6.7]): for all $\Psi \in \mathcal{F}_f(\mathcal{K})$,

$$(B(u)^*\Psi)^{(0)} = 0, \quad (B(u)^*\Psi)^{(p)} = \sqrt{p}A_p(u \otimes \Psi^{(p-1)}), \quad p \in \mathbb{N}.$$

The operator $B(u)$ is called the **fermion annihilation operator** with test vector u, while $B(u)^*$ is called the **fermion creation operator** with test vector u. The operator set $\{B(u), B(u)^* | u \in \mathcal{K}\}$ obeys the canonical anti-commutation relations (CAR): for all $u, v \in \mathcal{K}$,

$$\{B(u), B(v)^*\} = \langle u, v \rangle, \tag{2.17}$$

$$\{B(u), B(v)\} = 0, \quad \{B(u)^*, B(v)^*\} = 0. \tag{2.18}$$

Taking $v = u$ in (2.18), we have

$$B(u)^2 = 0, \quad (B(u)^*)^2 = 0, \quad u \in \mathcal{K}. \tag{2.19}$$

The Fock vacuum $\Omega_{\mathcal{K}} \in \mathcal{F}(\mathcal{K})$ belongs also to $\mathcal{F}_f(\mathcal{K})$ and satisfies

$$B(u)\Omega_{\mathcal{K}} = 0, \quad u \in \mathcal{K}. \tag{2.20}$$

Using (2.17), one can show that the operator norm $\|B(u)^\#\|$ of $B(u)^\#$ is given by

$$\|B(u)\| = \|B(u)^*\| = \|u\|. \tag{2.21}$$

In the same way as in the proof of (2.12), one can show that, for each subspace $\mathcal{E}$ of $\mathcal{K}$, we have

$$\mathcal{F}_{f,\mathrm{fin}}(\mathcal{E}) = \mathrm{span}\left\{\Omega_{\mathcal{K}}, \left(\prod_{j=1}^p B(u_j)^*\right)\Omega_{\mathcal{K}} \,|\, p \geq 1, u_j \in \mathcal{E}, j = 1, \ldots, p\right\}.$$

Let T be a densely defined closed linear operator on $\mathcal{K}$. Then, it follows that $\mathcal{F}_{f,\mathrm{fin}}(\mathrm{Dom}(T)) \subset \mathrm{Dom}(d\Gamma_f(T))$ and, for all $p \geq 1$ and $u_j \in \mathrm{Dom}(T)$, $j = 1, \ldots, p$,

$$d\Gamma_f(T)B(u_1)^* \cdots B(u_p)^*\Omega_{\mathcal{K}} = \sum_{j=1}^p B(u_1)^* \cdots B(Tu_j)^* \cdots B(u_p)^*\Omega_{\mathcal{K}}. \tag{2.22}$$

This formula and (2.21) imply:

Lemma 2.2 *Assume that $T \in \mathfrak{B}(\mathcal{K})$. Then, for all $\Psi \in \mathscr{F}_{\mathrm{f,fin}}(\mathcal{K})$,*

$$\|d\Gamma_{\mathrm{f}}(T)\Psi\| \leq C_{\Psi}\|T\|,$$

where C_{Ψ} is a constant depending on Ψ (independent of T).

Let T be a self-adjoint operator on $\mathcal{K}$. Then it is shown [22, Theorem 6.18] that, for all $t \in \mathbb{R}$ and $u \in \mathcal{K}$,

$$e^{itd\Gamma_{\mathrm{f}}(T)} B(u)^{\#} e^{-itd\Gamma_{\mathrm{f}}(T)} = B(e^{itT})^{\#}. \tag{2.23}$$

2.12 Fermion Quadratic Operators

For later use (see Sect. 4.11), we here recall some basic objects in the operator theory on the fermion Fock space $\mathscr{F}_{\mathrm{f}}(\mathcal{K})$.

Let $\{e_n\}_{n=1}^{\infty}$ be a complete orthonormal system (CONS) of $\mathcal{K}$.

Lemma 2.3 *Let $T \in \mathfrak{B}(\mathcal{K})$. Then, for all $\Psi \in \mathscr{F}_{\mathrm{f,fin}}(\mathcal{K})$,*

$$\lim_{N \to \infty} \sum_{n=1}^{N} B(Te_n)^* B(e_n)\Psi = d\Gamma_{\mathrm{f}}(T)\Psi, \tag{2.24}$$

$$\lim_{N \to \infty} \sum_{n=1}^{N} B(e_n)^* B(Te_n)\Psi = d\Gamma_{\mathrm{f}}(T^*)\Psi, \tag{2.25}$$

in $\mathscr{F}_{\mathrm{f}}(\mathcal{K})$ independently of the choice of $\{e_n\}_{n=1}^{\infty}$.

Proof It is sufficient to prove (2.24) and (2.25) for vectors Ψ of the form

$$\Psi = B(u_1)^* \cdots B(u_p)^* \Omega_{\mathcal{K}} \quad (u_1, \ldots, u_p \in \mathcal{K}). \tag{2.26}$$

In this case, we have

$$\sum_{n=1}^{N} B(Te_n)^* B(e_n)\Psi = \sum_{j=1}^{p} B(u_1)^* \cdots B(Tu_j^{(N)})^* \cdots B(u_p)^* \Omega_{\mathcal{K}},$$

where $u_j^{(N)} := \sum_{n=1}^{N} \langle e_n, u_j \rangle e_n$. It is obvious that $\lim_{N \to \infty} u_j^{(N)} = u_j$. Then the boundedness of T implies that $\lim_{N \to \infty} Tu_j^{(N)} = Tu_j$. Hence, by (2.21), we obtain

$$\lim_{N \to \infty} \sum_{n=1}^{N} B(Te_n)^* B(e_n) \Psi = \sum_{j=1}^{n} B(u_1)^* \cdots B(Tu_j)^* \cdots B(u_p)^* \Omega_{\mathscr{K}} = d\Gamma_{\mathrm{f}}(T)\Psi,$$

where we have used (2.22). Similarly, one can prove (2.25). □

A mapping $T : \mathscr{K} \to \mathscr{K}$ is called an anti-linear Hilbert–Schmidt operator if T is a bounded anti-linear operator and $\sum_{n=1}^{\infty} \|T\xi_n\|^2 < \infty$ for some CONS $\{\xi_n\}_{n=1}^{\infty}$ of $\mathscr{K}$. In this case, $\|T\|_2 := \sqrt{\sum_{n=1}^{\infty} \|T\xi_n\|^2}$ is called the Hilbert–Schmidt norm of T as in the usual Hilbert–Schmidt operators.[1]

Lemma 2.4 *Assume that T is a Hilbert–Schmidt operator on $\mathscr{K}$ or an anti-linear Hilbert–Schmidt operator on $\mathscr{K}$. Then, for all $\Psi \in \mathscr{F}_{\mathrm{f,fin}}(\mathscr{K})$, the limits*

$$q_1(T)\Psi := \lim_{N \to \infty} \sum_{n=1}^{N} B(Te_n)^* B(e_n)^* \Psi,$$

$$q_2(T)\Psi := \lim_{N \to \infty} \sum_{n=1}^{N} B(e_n) B(Te_n) \Psi$$

exist independently of the choice of $\{e_n\}_{n=1}^{\infty}$. Moreover,

$$\|q_1(T)\Psi\| \le C_\Psi \|T\|_2, \tag{2.27}$$

$$\|q_2(T)\Psi\| \le C_\Psi \|T\|. \tag{2.28}$$

Proof Let Ψ be as in (2.26). Then

$$\left(\sum_{n=1}^{N} B(Te_n)^* B(e_n)^* \Psi \right)^{(p+2)} = \sqrt{(p+2)!} \, A_{p+2}(\theta_N \otimes u_1 \otimes \cdots \otimes u_p),$$

where $\theta_N := \sum_{n=1}^{N} Te_n \otimes e_n$. It is easy to see that $\{\theta_N\}_{N=1}^{\infty}$ is a Cauchy sequence in $\mathscr{K} \otimes \mathscr{K}$. Hence $\theta := \lim_{N \to \infty} \theta_N = \sum_{n=1}^{\infty} Te_n \otimes e_n \in \mathscr{K} \otimes \mathscr{K}$ exists. Since A_{p+2} is a bounded operator on $\otimes^{p+2} \mathscr{K}$, it follows that $q_1(T)\Psi$ exists and

$$(q_1(T)\Psi)^{(p+2)} = \sqrt{(p+2)!} \, A_{p+2}(\theta \otimes u_1 \otimes \cdots \otimes u_p).$$

Hence $\|q_1(T)\Psi\| \le \sqrt{(p+2)!} \|\theta\| \|u_1\| \cdots \|u_p\|$. It is easy to see that $\|\theta\| = \|T\|_2$. Therefore (2.27) holds.

Let $\Phi_p := (\sum_{n=1}^{N} B(e_n) B(Te_n) \Psi)^{(p-2)}$. Then we have

---

[1] It is shown that $\sum_{n=1}^{\infty} \|T\xi_n\|^2$ is independent of the choice of $\{\xi_n\}_{n=1}^{\infty}$.

$$\Phi_p = \sum_{j=1}^{p} \left\{ \sum_{k=1}^{j-1} (-1)^{j+k} \left\langle \theta_j^{(N)}, u_k \right\rangle u_1 \otimes \cdots \otimes \hat{u}_k \otimes \cdots \hat{u}_j \otimes \cdots \otimes u_p \right.$$

$$\left. + \sum_{k=j+1}^{p} (-1)^{j+k-1} \left\langle \theta_j^{(N)}, u_k \right\rangle u_1 \otimes \cdots \otimes \hat{u}_j \otimes \cdots \hat{u}_k \otimes \cdots \otimes u_p \right\}$$

with $\theta_j^{(N)} := \sum_{n=1}^{N} \left\langle u_j, T e_n \right\rangle e_n$, where $\hat{u}_k$ indicates the omission of u_k. It is easy to see that $\theta_j := \lim_{N \to \infty} \theta_j^{(N)}$ exists and $\|\theta_j\| == \|T^* u_j\|$. Therefore $q_2(T)$ exists and

$$(q_2(T)\Psi)^{(p-2)}$$

$$= \sum_{j=1}^{p} \left\{ \sum_{k=1}^{j-1} (-1)^{j+k} \left\langle \theta_j, u_k \right\rangle u_1 \otimes \cdots \otimes \hat{u}_k \otimes \cdots \hat{u}_j \otimes \cdots \otimes u_p \right.$$

$$\left. + \sum_{k=j+1}^{p} (-1)^{j+k-1} \left\langle \theta_j, u_k \right\rangle u_1 \otimes \cdots \otimes \hat{u}_j \otimes \cdots \hat{u}_k \otimes \cdots \otimes u_p \right\}.$$

Note that $\|\theta_j\| \leq \|T^*\| \|u_j\|$ and $\|T^*\| = \|T\|$.[2] Hence it follows that $\|q_2(T)\Psi\| \leq C_\Psi \|T\|$. Thus (2.28) holds. $\qquad \square$

We regard $q_a(T)$ $(a = 1, 2)$ as a linear operator on $\mathscr{F}_{\mathrm{f}}(\mathscr{K})$ with $\mathrm{Dom}(q_a(T)) = \mathscr{F}_{\mathrm{f,fin}}(\mathscr{K})$. Hence $q_a(T)$ is densely defined. It follows that $\mathrm{Dom}(q_a(T)^*)$ includes $\mathscr{F}_{\mathrm{f,fin}}(\mathscr{K})$ and

$$q_1(T)^* \upharpoonright \mathscr{F}_{\mathrm{f,fin}}(\mathscr{K}) = q_2(T), \quad q_2(T)^* \upharpoonright \mathscr{F}_{\mathrm{f,fin}}(\mathscr{K}) = q_1(T).$$

Hence each $q_a(T)$ is closable. We introduce the following symbols:

$$\left\langle B^* | T | B^* \right\rangle := \overline{q_1(T)}, \quad \left\langle B | T | B \right\rangle := \overline{q_2(T)}.$$

Then we have $\left\langle B | T | B \right\rangle^* \supset \left\langle B^* | T | B^* \right\rangle$. In view of Lemma 2.3, for each $T \in \mathfrak{B}(\mathscr{K})$, we introduce the following symbol:

$$\left\langle B^* | T | B \right\rangle := d\Gamma_{\mathrm{f}}(T).$$

Each of the operators $\left\langle B^* | T | B \right\rangle$, $\left\langle B^* | T | B^* \right\rangle$ and $\left\langle B | T | B \right\rangle$ is called a **fermion quadratic operator** with respect to T.

---

[2] This holds also for bounded anti-linear operators.

Chapter 3
Q-space Representation of Boson Fock Space

Abstract We review the so-called Q-space representation (a probability theoretical representation) of the boson Fock space over a Hilbert space. This representation is useful in analyzing quantum field models (e.g., [20, 34, 38, 61, 62]) and has important relations to infinite-dimensional stochastic analysis (e.g., [38, 51]).

3.1 Gaussian Random Process

Let $\mathfrak{h}$ be a real Hilbert space and (M, Σ, μ) be a probability measure space. Suppose that, for each $f \in \mathfrak{h}$, a random variable $\varphi(f)$ on (M, Σ, μ) is assigned. If the set $\{\varphi(f) | f \in \mathfrak{h}\}$ of random variables satisfies the following properties, then it is called the **Gaussian random process** indexed by $\mathfrak{h}$:

(i) For all $f, g \in \mathfrak{h}$ and $a, b \in \mathbb{R}$,

$$\varphi(af + bg) = a\varphi(f) + b\varphi(g), \quad \text{a.e.,}$$

where "a.e." means "almost everywhere with respect to μ".
(ii) $\{\varphi(f) | f \in \mathfrak{h}\}$ is full, i.e., Σ is the smallest Borel field such that $\{\varphi(f) | f \in \mathfrak{h}\}$ is measurable.
(iii) For each $f \in \mathfrak{h}$, $\varphi(f)$ is a Gaussian random variable such that its characteristic function $t \in \mathbb{R} \mapsto \int_M e^{it\varphi(f)} d\mu$ is of the form:

$$\int_M e^{it\varphi(f)} d\mu = e^{-t^2 \|f\|_{\mathfrak{h}}^2 / 4}, \quad t \in \mathbb{R} \tag{3.1}$$

The argument of $\varphi(f)$ (i.e., points of M) will not be written explicitly if there is no danger of confusion.

Let $\{\varphi(f) | f \in \mathfrak{h}\}$ be the Gaussian random process indexed by $\mathfrak{h}$ and (M, Σ, μ) be the underlying probability measure space. It follows from (3.1) that, for all $p \in \mathbb{N}$, $\varphi(f)^p$ is in $L^2(M, f\mu)$ and

$$\langle \varphi(f)\varphi(g)\rangle = \frac{1}{2}\langle f, g\rangle_{\mathfrak{h}}, \quad f, g \in \mathfrak{h},$$

where, for an integrable function F on (M, Σ, μ), $\langle F\rangle := \int_M F \, d\mu$.

For each $n \in \mathbb{N}$ and $f_1, \ldots, f_n \in \mathfrak{h}$, one can define a random variable $: \varphi(f_1) \cdots \varphi(f_n) :$ on M by the following recursion relations:

$$: \varphi(f_1) : = \varphi(f_1),$$
$$: \varphi(f_1) \cdots \varphi(f_n) : = \varphi(f_1) : \varphi(f_2) \cdots \varphi(f_n) :$$
$$- \sum_{j=2}^{n} \langle \varphi(f_1)\varphi(f_j)\rangle : \varphi(f_2) \cdots \widehat{\varphi(f_j)} \cdots \varphi(f_n) :, \quad n \geq 2,$$

where $\widehat{\varphi(f_j)}$ indicates the omission of $\varphi(f_j)$. The random variable $: \varphi(f_1) \cdots \varphi(f_n) :$ is called the **Wick product** of $\varphi(f_1) \cdots \varphi(f_n)$.

For each $f \in \mathfrak{h}$ and $n \in \mathbb{N}$, we define $: \varphi(f)^n :$ by

$$: \varphi(f)^n : =: \underbrace{\varphi(f) \cdots \varphi(f)}_{n} : .$$

It follows that

$$: \varphi(f)^n := \sum_{m=0}^{[n/2]} \frac{n!}{(n-2m)!m!} \left(-\frac{1}{2}\langle \varphi(f)^2\rangle\right)^m \varphi(f)^{n-2m},$$

where $[n/2]$ denotes the maximal integer not exceeding $n/2$. It is shown [22, Theorem 5.23] that $: \varphi(f_1) \cdots \varphi(f_n) :$ is symmetric for all permutations of $(f_1, \ldots, f_n)$ and, for all $n, m \in \mathbb{N}$ and $f_j, g_k, f \in \mathfrak{h}$ $(j = 1, \ldots, n, k = 1, \ldots, m)$,

$$\langle : \varphi(f_1) \cdots \varphi(f_n) : \rangle = 0,$$
$$\langle : \varphi(f_1) \cdots \varphi(f_n) :: \varphi(g_1) \cdots \varphi(g_m) : \rangle = \frac{\delta_{mn}}{2^n} \sum_{\sigma \in \mathfrak{S}_n} \langle f_1, g_{\sigma(1)}\rangle_{\mathfrak{h}} \cdots \langle f_n, g_{\sigma(n)}\rangle_{\mathfrak{h}},$$
$$\langle : \varphi(f)^n :^2\rangle = \frac{n!\|f\|_{\mathfrak{h}}^{2n}}{2^n}. \tag{3.2}$$

In particular, for $n \neq m$, $: \varphi(f_1) \cdots \varphi(f_n) :$ is orthogonal to $: \varphi(g_1) \cdots \varphi(g_m) :$. The constant function 1 is in $L^2(M, d\mu)$. Hence, introducing the closed subspaces

$$\Gamma_0(\mathfrak{h}) := \{\alpha 1 | \alpha \in \mathbb{C}\},$$
$$\Gamma_n(\mathfrak{h}) := \overline{\text{span}} \{: \varphi(f_1) \cdots \varphi(f_n) : | f_1, \ldots, f_n \in \mathfrak{h}\}, \quad n \geq 1$$

in $L^2(M, d\mu)$, where, for a subset $\mathscr{D}$ of a Hilbert space $\mathscr{K}$, $\overline{\mathscr{D}}$ denotes the closure of $\mathscr{D}$ in $\mathscr{K}$, we have a family $\{\Gamma_n(\mathfrak{h})\}_{n=0}^{\infty}$ of mutually orthogonal closed subspaces in $L^2(M, d\mu)$. An important fact is:

$$L^2(M, d\mu) = \oplus_{n=0}^{\infty} \Gamma_n(\mathfrak{h}),$$

the **Itô–Segal–Wiener decomposition** (for a proof, see, e.g., [22, Theorem 5.52], [61, (I.25)]).

3.2 Natural Isomorphism of Boson Fock Spaces

It is a well-known fact that, for any real Hilbert space $\mathfrak{h}$, the Gaussian random process indexed by $\mathfrak{h}$ exists (see, e.g., [61, Theorem I.9]). We denote the Gaussian random process and the underlying probability measure space by $\{\varphi_{\mathfrak{h}}(f)|f \in \mathfrak{h}\}$ and $(Q_{\mathfrak{h}}, \Sigma_{\mathfrak{h}}, \mu_{\mathfrak{h}})$ respectively. Hence

$$\int_{Q_{\mathfrak{h}}} e^{i\varphi_{\mathfrak{h}}(f)} d\mu_{\mathfrak{h}} = e^{-\|f\|_{\mathfrak{h}}^2/4}, \quad f \in \mathfrak{h}.$$

We denote the complexification of $\mathfrak{h}$ by $\mathfrak{h}_{\mathbb{C}}$ and the complex conjugation on $\mathfrak{h}_{\mathbb{C}}$ by C. Then one can consider the boson Fock space $\mathscr{F}_{\mathrm{b}}(\mathfrak{h}_{\mathbb{C}})$ over $\mathfrak{h}_{\mathbb{C}}$. We introduce the following operators:

$$\phi_C(f) := \Phi_S(f), \quad \pi_C(f) := \Phi_S(if), \quad f \in \mathfrak{h}.$$

By (2.15), $\{\phi_C(f), \pi_C(f)|f \in \mathfrak{h}\}$ satisfies the Heisenberg CCR on $\mathscr{F}_{\mathrm{b},0}(\mathfrak{h}_{\mathbb{C}})$:

$$[\phi_C(f), \pi_C(g)] = i \langle f, g \rangle,$$
$$[\phi_C(f), \phi_C(g)] = 0, \quad [\pi_C(f), \pi_C(g)] = 0, \quad f, g \in \mathfrak{h}.$$

Hence, for each subspace $\mathscr{W}$ of $\mathfrak{h}$, $(\mathscr{F}_{\mathrm{b}}(\mathfrak{h}_{\mathbb{C}}), \mathscr{F}_{\mathrm{b},0}(\mathfrak{h}_{\mathbb{C}}), \{\phi_C(f), \pi_C(f)|f \in \mathscr{W}\})$ is a representation of the Heisenberg CCR over $\mathscr{W}$ [23, §8.8]. It is called the **Fock representation of the Heisenberg CCR over $\mathscr{W}$**.

There is a natural isomorphism between $\mathscr{F}_{\mathrm{b}}(\mathfrak{h}_{\mathbb{C}})$ and $L^2(Q_{\mathfrak{h}}, d\mu_{\mathfrak{h}})$:

Theorem 3.1 *There exists a unitary operator U_{b} from $\mathscr{F}_{\mathrm{b}}(\mathfrak{h}_{\mathbb{C}})$ to $L^2(Q_{\mathfrak{h}}, d\mu_{\mathfrak{h}})$ such that $U_{\mathrm{b}}\Omega_{\mathfrak{h}_{\mathbb{C}}} = 1$ and, for all $n \in \mathbb{N}$ and $f_1, \ldots, f_n \in \mathfrak{h}$,*

$$U_{\mathrm{b}}A(f_1)^* \cdots A(f_n)^*\Omega_{\mathfrak{h}_{\mathbb{C}}} = 2^{n/2} : \varphi_{\mathfrak{h}}(f_1) \cdots \varphi_{\mathfrak{h}}(f_n) : .$$

Moreover, the following operator equality holds:

$$U_\mathfrak{b}\phi_C(f)U_\mathfrak{b}^{-1} = \varphi_\mathfrak{h}(f), \quad f \in \mathfrak{h}.$$

Proof See, e.g., [61, Theorem I.11] or [22, Theorem 5.53]. □

We call the unitary operator $U_\mathfrak{b}$ in Theorem 3.1 the **natural isomorphism** from $\mathscr{F}_\mathfrak{b}(\mathfrak{h}_\mathbb{C})$ to $L^2(Q_\mathfrak{h}, d\mu_\mathfrak{h})$. The Hilbert space $L^2(Q_\mathfrak{h}, d\mu_\mathfrak{h})$ is called the Q-**space representation** of the boson Fock space $\mathscr{F}_\mathfrak{b}(\mathfrak{h}_\mathbb{C})$. One of the advantages of Q-space representation is in that the quantum field $\phi_C(f)$ is represented as the multiplication operator by the function $\varphi_\mathfrak{h}(f)$ on $Q_\mathfrak{h}$.

For each $f \in \mathfrak{h}_\mathbb{C}$, we define

$$D_f := \sqrt{2}U_\mathfrak{b}A(f)U_\mathfrak{b}^{-1}. \tag{3.3}$$

Since $A(f)$ is closed and anti-linear in $f \in \mathfrak{h}_\mathbb{C}$, D_f is a closed linear operator on $L^2(Q_\mathfrak{h}, d\mu_\mathfrak{h})$ and anti-linear in f.

For each $n \in \mathbb{N}$, we denote by $\mathscr{P}_n$ the set of complex polynomials of n variables $z_1, \ldots, z_n$. One can show that, for each $n \in \mathbb{N}$ and all $P \in \mathscr{P}_n$, $P(\varphi_\mathfrak{h}(f_1), \ldots, \varphi_\mathfrak{h}(f_n))$ is in $\mathrm{Dom}(D_f)$ and

$$D_f P(\varphi_\mathfrak{h}(f_1), \ldots, \varphi_\mathfrak{h}(f_n)) = \sum_{j=1}^{n} \langle f, f_j \rangle (\partial_j P)(\varphi_\mathfrak{h}(f_1), \ldots, \varphi_\mathfrak{h}(f_n)),$$

where $(\partial_j P)(z_1, \ldots, z_n) := \partial P(z_1, \ldots, z_n)/\partial z_j$. Based on this fact, we call D_f the **directional functional differential operator** in f. For a subspace $\mathscr{D}$ of $\mathfrak{h}$, we define

$$\mathscr{P}(\mathscr{D}) := \mathrm{span}\,\{P(\varphi_\mathfrak{h}(f_1), \ldots, \varphi_\mathfrak{h}(f_n)) | n \in \mathbb{N}, P \in \mathscr{P}_n, f_1, \ldots, f_n \in \mathscr{D}\}.$$

Let $f \in \mathfrak{h}$. Then $\pi_C(f) = i\phi_C(f) - i\sqrt{2}A(f)$ on $\mathrm{Dom}(A(f)) \cap \mathrm{Dom}(A(f)^*)$. Hence, letting

$$\pi_\mathfrak{h}(f) := U_\mathfrak{b}\pi_C(f)U_\mathfrak{b}^{-1},$$

we have

$$\pi_\mathfrak{h}(f) = -iD_f + i\varphi_\mathfrak{h}(f)$$

on $\mathscr{P}(\mathfrak{h})$.

The range of test vectors of $\phi_C(\cdot)$ can be extended in a natural way to $\mathfrak{h}_\mathbb{C}$: for each $f = f_1 + if_2$ ($f_1, f_2 \in \mathfrak{h}$), we define

$$\phi_C(f) := \phi_C(f_1) + i\phi_C(f_2).$$

It follows that the correspondence $f \mapsto \phi_C(f)$ is complex linear on $\mathscr{F}_{b,0}(\mathfrak{h}_C)$. Then we have

$$U_b \phi_C(f) U_b^{-1} = \varphi_{\mathfrak{h}}(f), \quad f \in \mathfrak{h}_C,$$

where

$$\varphi_{\mathfrak{h}}(f_1 + i f_2) := \varphi_{\mathfrak{h}}(f_1) + i \varphi_{\mathfrak{h}}(f_2) \quad (f_1, f_2 \in \mathfrak{h}).$$

By (3.3), we have $D_f^* = \sqrt{2} U_b A(f)^* U_b^{-1}$ for all $f \in \mathfrak{h}_C$. On the other hand, we have $A(f)^* = \sqrt{2}\phi_C(f) - A(Cf)$ on

$$\mathscr{E}_f := \mathrm{Dom}(A(f_1)) \cap \mathrm{Dom}(A(f_2)) \cap \mathrm{Dom}(A(f_1)^*) \cap \mathrm{Dom}(A(f_2)^*).$$

Hence

$$D_f^* = -D_{Cf} + 2\varphi_{\mathfrak{h}}(f) \quad \text{on } U_b \mathscr{E}_f. \tag{3.4}$$

3.3 Gradient Operator

We introduce a subspace of $L^2(Q_{\mathfrak{h}}, d\mu_{\mathfrak{h}})$:

$$C^1(Q_{\mathfrak{h}}) := \{\Psi \in \cap_{f \in \mathfrak{h}_C} \mathrm{Dom}(D_f) | \text{ for each } f \in \mathfrak{h}_C, \text{ the mapping } f \mapsto (D_f \Psi)(q)$$
$$\text{is continuous for a.e. } q \in Q_{\mathfrak{h}}\}.$$

It is easy to see that $\mathscr{P}(\mathfrak{h}) \subset C^1(Q_{\mathfrak{h}})$ with

$$D_f P(\varphi_{\mathfrak{h}}(f_1), \dots, \varphi_{\mathfrak{h}}(f_n)) = \left\langle f, \sum_{j=1}^{n} (\partial_j P)(\varphi_{\mathfrak{h}}(f_1), \dots, \varphi_{\mathfrak{h}}(f_n)) f_j \right\rangle_{\mathfrak{h}_C}$$

for all $P(\varphi_{\mathfrak{h}}(f_1), \dots, \varphi_{\mathfrak{h}}(f_n)) \in \mathscr{P}(\mathfrak{h})$. In particular, $C^1(Q_{\mathfrak{h}})$ is dense in $L^2(Q_{\mathfrak{h}}, d\mu_{\mathfrak{h}})$. By the Riesz theorem, for each $\Psi \in C^1(Q_{\mathfrak{h}})$ and a.e. $q \in Q_{\mathfrak{h}}$, there exists a unique vector $g_\Psi(q) \in \mathfrak{h}_C$ such that

$$(D_f \Psi)(q) = \langle f, g_\Psi(q) \rangle_{\mathfrak{h}_C}, \quad f \in \mathfrak{h}_C.$$

In general, for a measure space (M, Σ, ν) (ν is not necessarily a probability measure) and a separable Hilbert space $\mathscr{K}$, we denote by $L^p(M, d\nu; \mathscr{K})$ ($p \geq 1$) the space of $\mathscr{K}$-valued L^p-functions on (M, Σ, ν) (cf. [55, §II.1, Example 6]):

$$L^p(M, dv; \mathcal{K}) := \{F : M \to \mathcal{K}, \ \Sigma\text{-measurable} | \int_M \|F(x)\|_{\mathcal{K}}^p \, dv(x) < \infty\}.$$

$$(3.5)$$

In the case $p = 2$, $L^2(M, dv; \mathcal{K})$ becomes a Hilbert space with inner product

$$\langle F, G \rangle := \int_M \langle F(x), G(x) \rangle_{\mathcal{K}} \, dv(x), \quad F, G \in L^2(M, dv; \mathcal{K}).$$

One can define a linear operator ∇ from $L^2(Q_{\mathfrak{h}}, d\mu_{\mathfrak{h}})$ to $L^2(Q_{\mathfrak{h}}, d\mu_{\mathfrak{h}}; \mathfrak{h}_{\mathbb{C}})$ as follows:

$$\mathrm{Dom}(\nabla) := \{\Psi \in C^1(Q_{\mathfrak{h}}) | g_\Psi \in L^2(Q_{\mathfrak{h}}, d\mu_{\mathfrak{h}}; \mathfrak{h}_{\mathbb{C}})\}, \qquad (3.6)$$

$$\nabla \Psi := g_\Psi, \quad \Psi \in \mathrm{Dom}(\nabla) \qquad (3.7)$$

so that

$$D_f \Psi = \langle f, \nabla \Psi \rangle_{\mathfrak{h}_{\mathbb{C}}}, \quad f \in \mathfrak{h}_{\mathbb{C}}.$$

We call ∇ the **gradient operator** on $L^2(Q_{\mathfrak{h}}, d\mu_{\mathfrak{h}})$. It follows from the closedness of D_f that ∇ is closed.

Remark 3.1 In the context of the theory of the abstract Wiener space, the operator ∇ corresponds to the H-differential operator D [35, 59].

It is easy to see that $\mathscr{P}(\mathfrak{h}) \subset \mathrm{Dom}(\nabla)$ with

$$\nabla P(\varphi_{\mathfrak{h}}(f_1), \ldots, \varphi_{\mathfrak{h}}(f_n)) = \sum_{j=1}^n (\partial_j P)(\varphi_{\mathfrak{h}}(f_1), \ldots, \varphi_{\mathfrak{h}}(f_n)) f_j$$

for all $P(\varphi_{\mathfrak{h}}(f_1), \ldots, \varphi_{\mathfrak{h}}(f_n)) \in \mathscr{P}(\mathfrak{h})$. Hence ∇ is densely defined. Thus ∇ is a densely defined closed linear operator. Therefore, by a general theorem, the adjoint ∇^* exists as a linear operator from $L^2(Q_{\mathfrak{h}}, d\mu_{\mathfrak{h}}; \mathfrak{h}_{\mathbb{C}})$ to $L^2(Q_{\mathfrak{h}}, d\mu_{\mathfrak{h}})$ and is densely defined. It is easy to see that $\mathscr{P}(\mathfrak{h}) \hat{\otimes} \mathfrak{h}_{\mathbb{C}} \subset \mathrm{Dom}(\nabla^*)$.

3.4 More General Natural Isomorphisms of Boson Fock Spaces

Let $\mathfrak{g}$ be a real Hilbert space and T be a unitary operator from $\mathfrak{g}$ to $\mathfrak{h}$. For each $f = f_1 + if_2 \in \mathfrak{g}_{\mathbb{C}}$ ($f_1, f_2 \in \mathfrak{g}$), we can extend T to the unitary operator from $\mathfrak{g}_{\mathbb{C}}$ to $\mathfrak{h}_{\mathbb{C}}$ by $Tf := Tf_1 + iTf_2$. Then, as we have seen in Sect. 2.10, $\Gamma_{\mathrm{b}}(T)$ is a unitary operator from $\mathscr{F}_{\mathrm{b}}(\mathfrak{g}_{\mathbb{C}})$ to $\mathscr{F}_{\mathrm{b}}(\mathfrak{h}_{\mathbb{C}})$ and satisfies

$$\Gamma_{\mathfrak{b}}(T) A_{\mathfrak{g}_{\mathbb{C}}}(f) \Gamma_{\mathfrak{b}}(T)^{-1} = A_{\mathfrak{h}_{\mathbb{C}}}(Tf), \quad f \in \mathfrak{g}_{\mathbb{C}}.$$

Let $U_{\mathfrak{b}} : \mathscr{F}_{\mathfrak{b}}(\mathfrak{h}_{\mathbb{C}}) \to L^2(Q_{\mathfrak{h}}, d\mu_{\mathfrak{h}})$ be the unitary operator in Theorem 3.1. Then

$$U_T := U_{\mathfrak{b}} \Gamma_{\mathfrak{b}}(T)$$

is a unitary operator from $\mathscr{F}_{\mathfrak{b}}(\mathfrak{g}_{\mathbb{C}})$ to $L^2(Q_{\mathfrak{h}}, d\mu_{\mathfrak{h}})$ satisfying

$$U_T \Omega_{\mathfrak{g}_{\mathbb{C}}} = 1,$$
$$U_T A_{\mathfrak{g}_{\mathbb{C}}}(f_1)^* \cdots A_{\mathfrak{g}_{\mathbb{C}}}(f_n)^* \Omega_{\mathfrak{g}_{\mathbb{C}}} = 2^{n/2} : \varphi_{\mathfrak{h}}(Tf_1) \cdots \varphi_{\mathfrak{h}}(Tf_n):,$$
$$n \geq 1, f_1, \ldots, f_n \in \mathfrak{g}.$$

We already know that $\mathscr{F}_{\mathfrak{b}}(\mathfrak{g}_{\mathbb{C}})$ is unitarily equivalent to $L^2(Q_{\mathfrak{g}}, d\mu_{\mathfrak{g}})$. Hence $L^2(Q_{\mathfrak{g}}, d\mu_{\mathfrak{g}})$ is unitarily equivalent to $L^2(Q_{\mathfrak{h}}, d\mu_{\mathfrak{h}})$ in such a way that, for all n and $f_1, \ldots, f_n \in \mathfrak{g}, : \varphi_{\mathfrak{g}}(f_1) \cdots \varphi_{\mathfrak{g}}(f_n) :$ corresponds to $: \varphi_{\mathfrak{h}}(Tf_1) \cdots \varphi_{\mathfrak{h}}(Tf_n) :$. We say that this type of unitary equivalence is natural. Hence, in this sense, $L^2(Q_{\mathfrak{h}}, d\mu_{\mathfrak{h}})$ also can be regarded as a Q-space representation of $\mathscr{F}_{\mathfrak{b}}(\mathfrak{g})$.

The isomorphism U_T is useful in applications to quantum field theory (see Sect. 5.2.2).

Chapter 4
Boson–Fermion Fock Spaces and Abstract Supersymmetric Quantum Field Models

Abstract We introduce some operators on the abstract boson–fermion Fock space, including exterior differential and Dirac operators, and describe fundamental properties of them. We present also abstract supersymmetric quantum field models.

4.1 Boson–Fermion Fock Space

Let $\mathscr{H}$ and $\mathscr{K}$ be Hilbert spaces. Then one can make the tensor product

$$\mathscr{F}(\mathscr{H},\mathscr{K}) := \mathscr{F}_{\mathrm{b}}(\mathscr{H}) \otimes \mathscr{F}_{\mathrm{f}}(\mathscr{K})$$

of the boson Fock space $\mathscr{F}_{\mathrm{b}}(\mathscr{H})$ and the fermion Fock space $\mathscr{F}_{\mathrm{f}}(\mathscr{K})$. This Hilbert space is called the **boson–fermion Fock space** over the pair $(\mathscr{H},\mathscr{K})$. For a subspace $\mathscr{D}$ of $\mathscr{H}$ and a subspace $\mathscr{E}$ of $\mathscr{K}$, we define

$$\mathscr{F}_{\mathrm{fin}}(\mathscr{D},\mathscr{E}) := \mathscr{F}_{\mathrm{b,fin}}(\mathscr{D}) \hat{\otimes} \mathscr{F}_{\mathrm{f,fin}}(\mathscr{E}).$$

It follows that, if $\mathscr{D}$ and $\mathscr{E}$ are dense in $\mathscr{H}$ and $\mathscr{K}$ respectively, then $\mathscr{F}_{\mathrm{fin}}(\mathscr{D},\mathscr{E})$ is dense in $\mathscr{F}(\mathscr{H},\mathscr{K})$.

We have by (2.1) the following natural identifications:

$$\mathscr{F}(\mathscr{H},\mathscr{K}) = \oplus_{p=0}^{\infty}\mathscr{F}^{(p)}(\mathscr{H},\mathscr{K}), \quad \mathscr{F}^{(p)}(\mathscr{H},\mathscr{K}) := \mathscr{F}_{\mathrm{b}}(\mathscr{H}) \otimes \wedge^{p}(\mathscr{K}).$$

Hence we have the orthogonal decomposition

$$\mathscr{F}(\mathscr{H},\mathscr{K}) = \mathscr{F}_{+}(\mathscr{H},\mathscr{K}) \oplus \mathscr{F}_{-}(\mathscr{H},\mathscr{K}) \tag{4.1}$$

with

$$\mathscr{F}_{+}(\mathscr{H},\mathscr{K}) := \oplus_{p=0}^{\infty}\mathscr{F}^{(2p)}(\mathscr{H},\mathscr{K}), \quad \mathscr{F}_{-}(\mathscr{H},\mathscr{K}) := \oplus_{p=0}^{\infty}\mathscr{F}^{(2p+1)}(\mathscr{H},\mathscr{K}).$$

A. Arai, *Infinite-Dimensional Dirac Operators and Supersymmetric Quantum Fields*, SpringerBriefs in Mathematical Physics, https://doi.org/10.1007/978-981-19-5678-2_4

Let N_f be the fermion number operator on $\mathscr{F}_f(\mathscr{K})$ (see (2.6) and (2.7)). Then

$$\Gamma_{bf} := e^{i\pi I \otimes N_f}$$

is a unitary self-adjoint operator on $\mathscr{F}(\mathscr{H}, \mathscr{K})$ satisfying $\Gamma_{bf} \upharpoonright \mathscr{F}_{\pm}(\mathscr{H}, \mathscr{K}) = \pm I$. We call Γ_{bf} the **grading operator** on the boson–fermion Fock space $\mathscr{F}(\mathscr{H}, \mathscr{K})$ with respect to the orthogonal decomposition (4.1).

There is another orthogonal decomposition of the boson–fermion Fock space. For each $r \in \mathbb{Z}_+$, we define a closed subspace

$$\mathscr{F}_r(\mathscr{H}, \mathscr{K}) := \oplus_{n+p=r}(\otimes_s^n \mathscr{H}) \otimes \wedge^p(\mathscr{K}). \tag{4.2}$$

Then we have

$$\mathscr{F}(\mathscr{H}, \mathscr{K}) = \oplus_{r=0}^{\infty} \mathscr{F}_r(\mathscr{H}, \mathscr{K}). \tag{4.3}$$

This orthogonal decomposition has the following meaning. The boson number operator N_b and the fermion number operator N_f yield a new operator

$$N_{tot} := N_b \otimes I + I \otimes N_f \tag{4.4}$$

on $\mathscr{F}(\mathscr{H}, \mathscr{K})$. Then it is easy to see that $\mathscr{F}_r(\mathscr{H}, \mathscr{K})$ is the eigenspace of N_{tot} with eigenvalue r. Namely, (4.3) is the direct sum of the eigenspaces of N_{tot}. We call N_{tot} the **total number operator** on $\mathscr{F}(\mathscr{H}, \mathscr{K})$.

4.2 Q-Space Representation of Boson–Fermion Fock Space

In what follows, we assume that $\mathscr{H}$ and $\mathscr{K}$ are separable (then $\mathscr{F}_b(\mathscr{H})$ and $\mathscr{F}_f(\mathscr{K})$ are separable). We fix a real Hilbert space $\mathfrak{h}$ such that $\mathscr{H} = \mathfrak{h}_{\mathbb{C}}$, the complexification of $\mathfrak{h}$. Let U_b be the unitary operator in Theorem 3.1. Then

$$\tilde{U}_b := U_b \otimes I$$

is a unitary operator from $\mathscr{F}(\mathscr{H}, \mathscr{K})$ to $L^2(Q_{\mathfrak{h}}, d\mu_{\mathfrak{h}}) \otimes \mathscr{F}_f(\mathscr{K})$. Hence, under the unitary operator $\tilde{U}_b$, the boson–fermion Fock space $\mathscr{F}(\mathscr{H}, \mathscr{K})$ is unitarily equivalent to $L^2(Q_{\mathfrak{h}}, d\mu_{\mathfrak{h}}) \otimes \mathscr{F}_f(\mathscr{K})$. Moreover, there exists a unitary operator U_{bf} from $L^2(Q_{\mathfrak{h}}, d\mu_{\mathfrak{h}}) \otimes \mathscr{F}_f(\mathscr{K})$ to $L^2(Q_{\mathfrak{h}}, d\mu_{\mathfrak{h}}; \mathscr{F}_f(\mathscr{K}))$ (see (3.5)) such that

$$U_{bf}(\Phi \otimes \Psi) = \Phi(\cdot)\Psi, \quad \Phi \in L^2(Q_{\mathfrak{h}}, d\mu_{\mathfrak{h}}), \Psi \in \mathscr{F}_f(\mathscr{K})$$

(see, e.g., [22, Theorem 2.6]). Hence the operator

$$V_{\mathrm{bf}} := U_{\mathrm{bf}} \tilde{U}_{\mathrm{b}} \tag{4.5}$$

is a unitary operator from $\mathscr{F}(\mathscr{H}, \mathscr{K})$ to

$$\mathfrak{F} := L^2(Q_{\mathfrak{h}}, d\mu_{\mathfrak{h}}; \mathscr{F}_{\mathrm{f}}(\mathscr{K})), \tag{4.6}$$

the Hilbert space of $\mathscr{F}_{\mathrm{f}}(\mathscr{K})$-valued L^2-functions on $(Q_{\mathfrak{h}}, \mu_{\mathfrak{h}})$. Therefore the boson–fermion Fock space $\mathscr{F}(\mathscr{H}, \mathscr{K})$ is unitarily equivalent to $\mathfrak{F}$. We call the Hilbert space $\mathfrak{F}$ the **Q-space representation of the boson–fermion Fock space** $\mathscr{F}(\mathscr{H}, \mathscr{K})$. It is easy to see that

$$\mathfrak{F} = \oplus_{p=0}^{\infty} L^2(Q_{\mathfrak{h}}, d\mu_{\mathfrak{h}}; \wedge^p(\mathscr{K})).$$

An element of $L^2(Q_{\mathfrak{h}}, d\mu_{\mathfrak{h}}; \wedge^p(\mathscr{K}))$ may be regarded as an L^2-differential form of order p on the space $Q_{\mathfrak{h}}$. Thus the boson–fermion Fock space $\mathscr{F}(\mathscr{H}, \mathscr{K})$ is unitarily equivalent to the infinite direct sum of Hilbert spaces consisting of L^2-differential forms on $Q_{\mathfrak{h}}$.

4.3 Exterior Differential Operators

4.3.1 Definitions and Basic Properties

We introduce a basic operator on $\mathscr{F}(\mathscr{H}, \mathscr{K})$. Let $S \in \mathfrak{C}(\mathscr{H}, \mathscr{K})$ such that the non-negative self-adjoint operator S^*S is reduced by $\mathfrak{h}$. Let $\{\xi_n\}_{n=1}^{\infty}$ be a CONS of $\mathscr{K}$ such that $\xi_n \in \mathrm{Dom}(S^*)$, $n \in \mathbb{N}$. Then, for each $N \in \mathbb{N}$, we define an operator $d_S^{(N)}$ by

$$d_S^{(N)} := \sum_{n=1}^{N} A(S^*\xi_n) \otimes B(\xi_n)^* \tag{4.7}$$

acting in $\mathscr{F}(\mathscr{H}, \mathscr{K})$. It is obvious that $\mathscr{F}_{\mathrm{b,fin}}(\mathscr{H}) \hat{\otimes} \mathscr{F}_{\mathrm{f}}(\mathscr{K}) \subset \mathrm{Dom}(d_S^{(N)})$. It follows from (2.11) that, for all $\phi \in \mathscr{F}_{\mathrm{f}}(\mathscr{K})$,

$$d_S^{(N)}(\Omega_{\mathscr{H}} \otimes \phi) = 0. \tag{4.8}$$

For $f_1, \ldots, f_n \in \mathscr{H}$ ($n \in \mathbb{N}$) and each $P \in \mathscr{P}_n$, we define vectors $\Psi(f_1, \ldots, f_n)$ and $\Psi_P(f_1, \ldots, f_n) \in \mathscr{F}_{\mathrm{b,fin}}(\mathscr{H})$ by

$$\Psi(f_1, \ldots, f_n) := A(f_1)^* \cdots A(f_n)^* \Omega_{\mathscr{H}},$$
$$\Psi_P(f_1, \ldots, f_n) := P(\phi_C(f_1), \ldots, \phi_C(f_n))\Omega_{\mathscr{H}}.$$

Lemma 4.1 *For all* $\Psi \in \mathscr{F}_{b, \mathrm{fin}}(\mathrm{Dom}(S)) \hat{\otimes} \mathscr{F}_f(\mathscr{K})$, *the limit*

$$d_S^{(\infty)} \Psi := \lim_{N \to \infty} d_S^{(N)} \Psi \tag{4.9}$$

exists and, for each $P \in \mathscr{P}_n$, *all* $f_1, \ldots, f_n \in \mathrm{Dom}(S)$ $(n \in \mathbb{N})$ *and all* $\phi \in \mathscr{F}_f(\mathscr{K})$,

$$d_S^{(\infty)}(\Omega_{\mathscr{H}} \otimes \phi) = 0, \tag{4.10}$$

$$d_S^{(\infty)}\Psi(f_1, \ldots, f_n) \otimes \phi = \sum_{j=1}^n \Psi(f_1, \ldots, \widehat{f_j}, \ldots, f_n) \otimes B(Sf_j)^* \phi, \tag{4.11}$$

$$d_S^{(\infty)}\Psi_P(f_1, \ldots, f_n) \otimes \phi = \sum_{j=1}^n \Psi_{\partial_j P}(f_1, \ldots, \widehat{f_j}, \ldots, f_n) \otimes B(Sf_j)^* \phi. \tag{4.12}$$

Proof Equation (4.10) follows from (4.8). Using (2.9) and (2.11), one can show that

$$d_S^{(N)}\Psi(f_1, \ldots, f_n) \otimes \phi = \sum_{j=1}^n \Psi(f_1, \ldots \widehat{f_j}, \ldots f_n) \otimes B(u_j^{(N)})^* \phi$$

with $u_j^{(N)} := \sum_{m=1}^N \langle \xi_m, Sf_j \rangle \xi_m$. We have $\lim_{N \to \infty} u_j^{(N)} = Sf_j$ in $\mathscr{K}$. Hence, by (2.21), $\lim_{N \to \infty} \|B(u_j^{(N)})^* - B(Sf_j)^*\| = 0$. Thus $d_S^{(\infty)}\Psi(f_1, \ldots, f_n) \otimes \phi)$ exists and (4.11) holds. Similarly, one can prove (4.12). $\qquad\square$

Equations (4.10) and (4.11) show that the limit operator $d_S^{(\infty)}$ is independent of the choice of the CONS $\{\xi_n\}_{n=1}^\infty$. Hence $d_S^{(\infty)}$ is well-defined with $\mathrm{Dom}(d_S^{(\infty)}) = \mathscr{F}_{b, \mathrm{fin}}(\mathrm{Dom}(S)) \hat{\otimes} \mathscr{F}_f(\mathscr{K})$.

Since $d_S^{(\infty)}$ is densely defined, its adjoint $(d_S^{(\infty)})^*$ exists. Basic properties of this operator are described in the next lemma:

Lemma 4.2 *It holds that*

$$\mathrm{Dom}((d_S^{(\infty)})^*) \supset \mathscr{F}_{\mathrm{fin}}(\mathscr{H}, \mathrm{Dom}(S^*))$$

and, for all $\psi \in \mathscr{F}_{b, \mathrm{fin}}(\mathscr{H})$, $u_1, \ldots, u_p \in \mathrm{Dom}(S^*)$ $(p \geq 1)$,

$$(d_S^{(\infty)})^*(\psi \otimes \Omega_{\mathscr{H}}) = 0, \tag{4.13}$$

$$(d_S^{(\infty)})^*(\psi \otimes B(u_1)^* \cdots B(u_p)^* \Omega_{\mathscr{H}})$$

$$= \sum_{k=1}^{p}(-1)^{k-1} A(S^* u_k)^* \psi \otimes B(u_1)^* \cdots \widehat{B(u_k)^*} \cdots B(u_p)^* \Omega_{\mathscr{H}}. \tag{4.14}$$

Proof One has $(d_S^{(N)})^* \supset \sum_{n=1}^{N} A(S^* \xi_n)^* \otimes B(\xi_n)$, which, together with (2.20), implies $(d_S^{(N)})^*(\psi \otimes \Omega_{\mathscr{H}}) = 0$. Hence, for all $\Phi \in \mathrm{Dom}(d_S^{(\infty)})$, $\langle \psi \otimes \Omega_{\mathscr{H}}, d_S^{(N)} \Phi \rangle = 0$. Taking the limit $N \to \infty$ and using Lemma 4.1, we have $\langle \psi \otimes \Omega_{\mathscr{H}}, d_S^{(\infty)} \Phi \rangle = 0$. This implies that $\psi \otimes \Omega_{\mathscr{H}} \in \mathrm{Dom}((d_S^{(\infty)})^*)$ and (4.13) holds.

We next prove (4.14). Let $a_N := \left\langle \psi \otimes B(u_1)^* \cdots B(u_p)^* \Omega_{\mathscr{H}}, d_S^{(N)} \Phi \right\rangle$. Then, using (2.17), we have

$$a_N = \left\langle (d_S^{(N)})^* \psi \otimes B(u_1)^* \cdots B(u_p)^* \Omega_{\mathscr{H}}, \Phi \right\rangle$$

$$= \sum_{k=1}^{p}(-1)^{k-1} \left\langle \psi \otimes B(u_1)^* \cdots \widehat{B(u_k)^*} \cdots B(u_p)^* \Omega_{\mathscr{H}}, (A(S^* f_k^{(N)}) \otimes I)\Phi \right\rangle,$$

where $f_k^{(N)} := \sum_{n=1}^{N} \langle \xi_n, u_k \rangle \xi_n$. By Lemma 4.1, we have

$$\lim_{N \to \infty} a_N = \left\langle \psi \otimes B(u_1)^* \cdots B(u_p)^* \Omega_{\mathscr{H}}, d_S^{(\infty)} \Phi \right\rangle.$$

We have $\lim_{N \to \infty} f_k^{(N)} = u_k$ in $\mathscr{H}$. Hence it follows that

$$\lim_{N \to \infty} \left\langle \psi \otimes B(u_1)^* \cdots \widehat{B(u_k)^*} \cdots B(u_p)^* \Omega_{\mathscr{H}}, (A(S^* f_k^{(N)}) \otimes I)\Phi \right\rangle$$

$$= \left\langle \psi \otimes B(u_1)^* \cdots \widehat{B(u_k)^*} \cdots B(u_p)^* \Omega_{\mathscr{H}}, (A(S^* u_k) \otimes I)\Phi \right\rangle$$

$$= \left\langle A(S^* u_k)\psi \otimes B(u_1)^* \cdots \widehat{B(u_k)^*} \cdots B(u_p)^* \Omega_{\mathscr{H}}, \Phi \right\rangle.$$

Hence

$$\left\langle \psi \otimes B(u_1)^* \cdots B(u_p)^* \Omega_{\mathscr{H}}, d_S^{(\infty)} \Phi \right\rangle$$

$$= \left\langle \sum_{k=1}^{p}(-1)^{k-1} A(S^* u_k)\psi \otimes B(u_1)^* \cdots \widehat{B(u_k)^*} \cdots B(u_p)^* \Omega_{\mathscr{H}}, \Phi \right\rangle.$$

This implies that $\psi \otimes B(u_1)^* \cdots B(u_p)^* \Omega_{\mathscr{H}} \in \mathrm{Dom}((d_S^{(\infty)})^*)$ and (4.14) holds. $\square$

By Lemma 4.2, the adjoint $(d_S^{(\infty)})^*$ is densely defined. Hence $d_S^{(\infty)}$ is closable. We denote its closure by d_S:

$$d_S := \overline{d_S^{(\infty)}}.\tag{4.15}$$

For a subspace $\mathscr{D}$ of $\mathscr{H}$ and $p \geq 0$, we define

$$\mathscr{V}_p(\mathscr{D}) := \text{span}\,\{\Psi_P(f_1,\ldots,f_n)|n \in \mathbb{N},\, f_1,\ldots,f_n \in \mathscr{D},\, P \in \mathscr{P}_n\}\hat{\otimes}\wedge^p(\mathscr{K}).$$

For vectors $X = \Psi_P(f_1,\ldots,f_n) \otimes \eta \in \mathscr{V}_p(\mathscr{D})$ and $Y = \Psi_Q(g_1,\ldots,g_m) \otimes \theta \in \mathscr{V}_q(\mathscr{D})$ $(p,q \geq 0)$, we define a vector in $\mathscr{V}_{p+q}(\mathscr{D})$ by

$$X \wedge Y := P(\phi_C(f_1),\ldots,\phi_C(f_n))Q(\phi_C(g_1),\ldots,\phi_C(g_m))\Omega_{\mathscr{H}} \otimes (\eta \wedge \theta),\tag{4.16}$$

where

$$\eta \wedge \theta := \frac{\sqrt{(p+q)!}}{\sqrt{p!q!}}A_{p+q}(\eta \otimes \theta) \in \wedge^{p+q}(\mathscr{K}),$$

the wedge (exterior) product of η and θ. For any $X \in \mathscr{V}_p(\mathscr{D})$ and $Y \in \mathscr{V}_q(\mathscr{D})$, we extend the operation $\wedge$ by bilinearity to define $X \wedge Y \in \mathscr{V}_{p+q}(\mathscr{D})$. We call $X \wedge Y$ the **wedge product** of X and Y.

The next theorem states basic properties of d_S:

Theorem 4.1

(i) *(nilpotency) For all* $\Psi \in \text{Dom}(d_S)$, $d_S\Psi \in \text{Dom}(d_S)$ *and* $d_S^2\Psi = 0$.
(ii) *For each* $p = 0,1,2,\cdots$, d_S *maps* $\text{Dom}(d_S) \cap \mathscr{F}^{(p)}(\mathscr{H},\mathscr{K})$ *to* $\mathscr{F}^{(p+1)}(\mathscr{H},\mathscr{K})$.
(iii) *For all* $X \in \mathscr{V}_p(\text{Dom}(S))$ *and* $Y \in \mathscr{V}_q(\text{Dom}(S))$, $X \wedge Y$ *is in* $\text{Dom}(d_S)$ *and*

$$d_S(X \wedge Y) = (d_S X) \wedge Y + (-1)^p X \wedge (d_S Y).\tag{4.17}$$

Proof (i)   Let $f_1,\ldots,f_n \in \text{Dom}(S)$, $\phi \in \mathscr{F}_f(\mathscr{K})$ and   set $\Psi :=$ $A(f_1)^* \cdots A(f_n)^*\Omega_{\mathscr{H}} \otimes \phi$. Then, using (2.10) and (2.18), we have

$$d_S^2\Psi = \sum_{j,k=1,j\neq k}^n A(f_1)^* \cdots \widehat{A(f_k)}^* \cdots \widehat{A(f_j)}^* \cdots A(f_n)^*\Omega_{\mathscr{H}} \otimes B(Sf_k)^*B(Sf_j)^*\phi$$

$$= -\sum_{j,k=1,j\neq k}^n A(f_1)^* \cdots \widehat{A(f_k)}^* \cdots \widehat{A(f_j)}^* \cdots A(f_n)^*\Omega_{\mathscr{H}} \otimes B(Sf_j)^*B(Sf_k)^*\phi$$

$$= -d_S^2\Psi.$$

Hence $d_S^2\Psi = 0$. Let $\Psi \in \text{Dom}(d_S)$. Then, by (4.15), there exists a sequence $\{\Psi_n\}_{n=1}^\infty$ with $\Psi_n \in \mathscr{F}_{\text{b,fin}}(\text{Dom}(S))\hat{\otimes}\mathscr{F}_f(\mathscr{K})$ such that $\Psi_n \to \Psi$ and $d_S\Psi_n \to d_S\Psi$ $(n \to \infty)$. By the preceding result, $d_S(d_S\Psi_n) = 0$. Since d_S is closed, it follows that $d_S\Psi \in \text{Dom}(d_S)$ and $d_S(d_S\Psi) = 0$.

(ii) This follows from (4.11) and a limiting argument.

(iii) It is sufficient to prove (4.17) for X and Y given by (4.16) with $f_1, \ldots, f_n$, $g_1, \ldots, g_m \in \mathrm{Dom}(S)$. In this case, we have

$$d_S(X \wedge Y) = \left(\sum_{j=1}^{n} \Psi_{\partial_j P}(f_1, \ldots, f_n) \right) \Psi_Q(g_1, \ldots, g_m) \otimes (Sf_j) \wedge (\eta \wedge \theta)$$

$$+ \Psi_P(f_1, \ldots, f_n) \left(\sum_{k=1}^{m} \Psi_{\partial_k Q}(g_1, \ldots, g_m) \right) \otimes (Sg_k) \wedge (\eta \wedge \theta).$$

It is easy to see that $(Sf_j) \wedge (\eta \wedge \theta) = (Sf_j \wedge \eta) \wedge \theta$ and $(Sg_k) \wedge (\eta \wedge \theta) = (-1)^p \eta \wedge (Sg_k \wedge \theta)$. Hence (4.17) follows. $\qquad \square$

Based on Theorem 4.1, we call the operator d_S the **exterior differential operator** associated with $S \in \mathfrak{C}(\mathcal{H}, \mathcal{K})$ on the boson–fermion Fock space $\mathcal{F}(\mathcal{H}, \mathcal{K})$. Anticommutation properties of the family $\{d_S\}_{S \in \mathfrak{C}(\mathcal{H}, \mathcal{K})}$ are summarized in the following lemma:

Lemma 4.3 *Let* $S, T \in \mathfrak{C}(\mathcal{H}, \mathcal{K})$. *Then:*

(i) $\mathcal{F}_{\mathrm{fin}}(\mathrm{Dom}(S) \cap \mathrm{Dom}(T), \mathcal{K}) \subset \mathrm{Dom}(d_S d_T) \cap \mathrm{Dom}(d_T d_S)$ *and*

$$\{d_S, d_T\} = 0 \quad \text{on } \mathcal{F}_{\mathrm{fin}}(\mathrm{Dom}(S) \cap \mathrm{Dom}(T), \mathcal{K}).$$

(ii) $\mathcal{F}_{\mathrm{fin}}(\mathcal{H}, \mathrm{Dom}(S^*) \cap \mathrm{Dom}(T^*)) \subset \mathrm{Dom}(d_S^* d_T^*) \cap \mathrm{Dom}(d_T^* d_S^*)$ *and*

$$\{d_S^*, d_T^*\} = 0 \quad \text{on } \mathcal{F}_{\mathrm{fin}}(\mathcal{H}, \mathrm{Dom}(S^*) \cap \mathrm{Dom}(T^*)).$$

(iii) $\mathcal{F}_{\mathrm{fin}}(\mathrm{Dom}(T^* S), \mathrm{Dom}(ST^*)) \subset \mathrm{Dom}(d_S d_T^*) \cap \mathrm{Dom}(d_T^* d_S)$ *and*

$$\{d_S, d_T^*\} = d\Gamma_{\mathrm{b}}^{(\mathrm{alg})}(T^* S) \hat{\otimes} I + I \hat{\otimes} d\Gamma_{\mathrm{f}}^{(\mathrm{alg})}(ST^*)$$

on $\mathcal{F}_{\mathrm{fin}}(\mathrm{Dom}(T^* S), \mathrm{Dom}(ST^*))$, *where* $d\Gamma_{\mathrm{b}}^{(\mathrm{alg})}(\#)$ *(resp.* $d\Gamma_{\mathrm{f}}^{(\mathrm{alg})}(\#)$*) is the reduction of the algebraic second quantization* $d\Gamma^{(\mathrm{alg})}(\#)$ *of* $\#$ *(see (2.5)) to the boson (resp. fermion) Fock space.*

Proof Direct computations as in the proofs of Lemmas 4.1 and 4.2 (use (2.13) and (2.22) also). $\qquad \square$

Remark 4.1 Miyao [53] gave a characterization for the operator d_S from a more general point of view, called the *super-quantization*.

4.3.2 A Cochain Complex

By Theorem 4.1(ii), one can define for each $p \geq 0$ a densely defined closed linear operator $d_{S,p}$ from $\mathscr{F}^{(p)}(\mathscr{H}, \mathscr{K})$ to $\mathscr{F}^{(p+1)}(\mathscr{H}, \mathscr{K})$ by

$$\mathrm{Dom}(d_{S,p}) := \mathrm{Dom}(d_S) \cap \mathscr{F}^{(p)}(\mathscr{H}, \mathscr{K}),$$
$$d_{S,p}\Psi := (d_S\Psi)^{(p+1)}, \quad \Psi \in \mathrm{Dom}(d_p).$$

Then Theorem 4.1(ii) implies that

$$\mathrm{Ran}(d_{S,p}) \subset \mathrm{Dom}(d_{S,p+1}), \tag{4.18}$$
$$d_{S,p+1}d_{S,p} = 0 \quad \text{on } \mathrm{Dom}(d_{S,p}). \tag{4.19}$$

This shows that $(\{\mathrm{Dom}(d_{S,p})\}_{p=0}^{\infty}, \{d_{S,p}\}_{p=0}^{\infty})$ forms a cochain complex:

$$0 \to \mathrm{Dom}(d_{S,0}) \xrightarrow{d_{S,0}} \mathrm{Dom}(d_{S,1}) \xrightarrow{d_{S,1}} \cdots \to \mathrm{Dom}(d_{S,p}) \xrightarrow{d_{S,p}} \mathrm{Dom}(d_{S,p+1}) \to \cdots$$
$$\tag{4.20}$$

4.4 Operators in the Q-space Representation

Let V_{bf} be the unitary operator defined by (4.5). Then, each linear operator L on $\mathscr{F}(\mathscr{H}, \mathscr{K})$ acts in $L^2(Q_{\mathfrak{h}}, d\mu_{\mathfrak{h}}; \mathscr{F}_{\mathrm{f}}(\mathscr{K}))$ as $V_{\mathrm{bf}} L V_{\mathrm{bf}}^{-1}$. By (3.3), we have for all $f \in \mathscr{H}$

$$\tilde{U}_{\mathfrak{b}}(A(f) \otimes I)\tilde{U}_{\mathfrak{b}}^{-1} = \frac{1}{\sqrt{2}}D_f \otimes I.$$

Hence

$$V_{\mathrm{bf}}d_S^{(N)}V_{\mathrm{bf}}^{-1} = \sum_{n=1}^{N} \frac{1}{\sqrt{2}}B(\xi_n)^* D_{S^*\xi_n}.$$

Therefore

$$V_{\mathrm{bf}}d_S V_{\mathrm{bf}}^{-1} = \sum_{n=1}^{\infty} \frac{1}{\sqrt{2}}B(\xi_n)^* D_{S^*\xi_n} \quad \text{on } V_{\mathrm{bf}}\mathscr{F}_{\mathrm{fin}}(\mathrm{Dom}(S), \mathscr{K}). \tag{4.21}$$

Note that

$$\{B(\xi_n), B(\xi_m)^*\} = \delta_{nm}, \quad \{B(\xi_n), B(\xi_m)\} = 0, \quad n, m \in \mathbb{N}.$$

Hence $V_{\mathrm{bf}} d_S V_{\mathrm{bf}}^{-1}$ is a functional differential operator of infinite variables with bounded operator coefficients satisfying CAR.

4.5 Hilbert Complex

The structure shown in (4.20) with (4.18) and (4.19) is in fact an example of a general concept, called the Hilbert complex [30]. In this section, we review this concept and related facts.

4.5.1 Definitions and Basic Facts

Let $\{\mathscr{H}_p\}_{p=0}^{\infty}$ be a sequence of Hilbert spaces and $\{D_p\}_{p=0}^{\infty}$ be a sequence of densely defined closed linear operators with domain

$$\mathscr{D}_p := \mathrm{Dom}(D_p) \subset \mathscr{H}_p$$

and range

$$\mathscr{R}_p := \mathrm{Ran}(D_p) \subset \mathscr{H}_{p+1}$$

such that, for all $p \geq 0$, $\mathscr{R}_p \subset \mathscr{D}_{p+1}$ and

$$D_{p+1} D_p = 0 \quad \text{on } \mathscr{D}_p. \tag{4.22}$$

Hence we have a cochain complex

$$0 \longrightarrow \mathscr{D}_0 \xrightarrow{D_0} \mathscr{D}_1 \xrightarrow{D_1} \cdots \longrightarrow \mathscr{D}_p \xrightarrow{D_p} \mathscr{D}_{p+1} \longrightarrow \cdots$$

This cochain complex, denoted by $(\{\mathscr{H}_p\}_{p=0}^{\infty}, \{\mathscr{D}_p\}_{p=0}^{\infty}, \{D_p\}_{p=0}^{\infty})$, is called a **Hilbert complex**. We use the following convention:

$$\mathscr{H}_{-1} := \mathscr{D}_{-1} := \{0\}, \quad D_{-1} := 0. \tag{4.23}$$

If there is a number $N \in \mathbb{N}$ such that $\mathscr{H}_p = \{0\}, \forall p \geq N + 1$, then the Hilbert complex is said to be finite. A finite Hilbert complex has been discussed in [30]. In this section, we deal with the case where the Hilbert complex is infinite, i.e., the case where there exist infinitely many non-zero Hilbert spaces in $\{\mathscr{H}_p\}_{p=0}^{\infty}$.

It follows from (4.22) that, for all $p \geq 0$,

$$\mathscr{R}_p^* := \mathrm{Ran}(D_p^*) \subset \mathrm{Dom}(D_{p-1}^*) \subset \mathscr{H}_p$$

and

$$D_{p-1}^* D_p^* = 0 \quad \text{on } \mathscr{D}_p^* := \text{Dom}(D_p^*).$$

By (4.23), $\mathscr{D}_{-1}^* = \mathscr{H}_0$ and $D_{-1}^* = 0$. Hence we have a chain complex

$$0 \longleftarrow \mathscr{D}_{-1}^* \xleftarrow{D_0^*} \mathscr{D}_0^* \xleftarrow{D_1^*} \mathscr{D}_1^* \longleftarrow \cdots \mathscr{D}_{p-1}^* \xleftarrow{D_p^*} \mathscr{D}_p^* \longleftarrow \cdots .$$

We call the chain complex $(\{\mathscr{H}_p\}_{p=-1}^\infty, \{\mathscr{D}_p^*\}_{p=-1}^\infty, \{D_p^*\}_{p=-1}^\infty)$ the **dual complex** of the Hilbert complex $(\{\mathscr{H}_p\}_{p=0}^\infty, \{\mathscr{D}_p\}_{p=0}^\infty, \{D_p\}_{p=0}^\infty)$.

In every Hilbert complex $(\{\mathscr{H}_p\}_{p=0}^\infty, \{\mathscr{D}_p\}_{p=0}^\infty, \{D_p\}_{p=0}^\infty)$, each Hilbert space $\mathscr{H}_p$ has a natural orthogonal decomposition as shown below. Let

$$\hat{\mathscr{H}}_p := \ker D_p \cap \ker D_{p-1}^*, \quad p \geq 0.$$

Then one has the following fact [30, Lemma 2.1]:

Theorem 4.2 (weak Hodge decomposition) *For each $p \geq 0$, $\mathscr{H}_p$ has the orthogonal decomposition*

$$\mathscr{H}_p = \hat{\mathscr{H}}_p \oplus \overline{\mathscr{R}_{p-1}} \oplus \overline{\mathscr{R}_p^*}. \tag{4.24}$$

Proof Since D_p is a closed linear operator, $\ker D_p$ is a closed subspace of $\mathscr{H}_p$. Hence one has the orthogonal decomposition

$$\mathscr{H}_p = \ker D_p \oplus (\ker D_p)^\perp,$$

where $(\ker D_p)^\perp$ denotes the orthogonal complement of $\ker D_p$. The closedness of $\ker D_p$ and (4.22) imply that $\overline{\mathscr{R}_{p-1}} \subset \ker D_p$. One has also $\mathscr{H}_p = \overline{\mathscr{R}_{p-1}} \oplus \mathscr{R}_{p-1}^\perp$. Hence $\ker D_p = \overline{\mathscr{R}_{p-1}} \oplus (\ker D_p \cap \mathscr{R}_{p-1}^\perp)$. Therefore

$$\mathscr{H}_p = \overline{\mathscr{R}_{p-1}} \oplus (\ker D_p \cap \mathscr{R}_{p-1}^\perp) \oplus (\ker D_p)^\perp.$$

On the other hand, for each closed linear operator T on a Hilbert space, $(\ker T)^\perp = \overline{\text{Ran}(T^*)}$ (e.g., [23, Theorem 1.2]), which implies also that $(\text{Ran} T)^\perp = \ker T^*$. Therefore

$$\mathscr{H}_p = \overline{\mathscr{R}_{p-1}} \oplus (\ker D_p \cap \ker D_{p-1}^*) \oplus \overline{\mathscr{R}_p^*}.$$

Thus (4.24) holds. □

4.5.2 Laplace–Beltrami Operators of Finite Order and de Rham–Hodge–Kodaira Decompositions

By the von Neumann theorem, for each $p \geq 0$, $D_p^* D_p$ and $D_p D_p^*$ are non-negative self-adjoint operators on $\mathscr{H}_p$ and $\mathscr{H}_{p+1}$ respectively. We set

$$\Delta_0 := D_0^* D_0.$$

In what follows, we assume the following:

Assumption (A) *For each $p \geq 1$, the subspace*

$$\mathscr{E}_p := \mathscr{D}_p \cap \mathscr{D}_{p-1}^*$$

is dense in $\mathscr{H}_p$.

Lemma 4.4 *For each $p \geq 1$, there exists a unique non-negative self-adjoint operator Δ_p on $\mathscr{H}_p$ such that $\mathrm{Dom}(\Delta_p^{1/2}) = \mathscr{E}_p$ and, for all $\Psi, \Phi \in \mathrm{Dom}(\Delta_p^{1/2})$,*

$$\langle \Delta_p^{1/2}\Psi, \Delta_p^{1/2}\Phi \rangle = \langle D_p\Psi, D_p\Phi \rangle + \langle D_{p-1}^*\Psi, D_{p-1}^*\Phi \rangle. \tag{4.25}$$

Proof It is easy to see that the sesquilinear form q on $\mathscr{E}_p$ defined by

$$q(\Psi, \Phi) := \langle D_p\Psi, D_p\Phi \rangle + \langle D_{p-1}^*\Psi, D_{p-1}^*\Phi \rangle, \quad \Psi, \Phi \in \mathscr{E}_p$$

is closed and non-negative. Hence, by a general representation theorem on sesquilinear forms (e.g., [45, Chap. VI, Theorem 2.23], [55, Theorem VIII.15]), the operator Δ_p as described above exists. $\qquad \square$

It follows from (4.25) that, for all $p \geq 1$, $\mathrm{Dom}(D_p^* D_p) \cap \mathrm{Dom}(D_{p-1} D_{p-1}^*) \subset \mathrm{Dom}(\Delta_p)$ and

$$\Delta_p \Psi = (D_p^* D_p + D_{p-1} D_{p-1}^*)\Psi, \quad \Psi \in \mathrm{Dom}(D_p^* D_p) \cap \mathrm{Dom}(D_{p-1} D_{p-1}^*).$$

In other words,

$$D_p^* D_p + D_{p-1} D_{p-1}^* \subset \Delta_p. \tag{4.26}$$

We remark that, at this stage, it is unclear whether the equality (operator equality) holds in (4.26) or not (see Remark 4.2). By this fact and by analogy with the Laplace–Beltrami operator on the pth differential forms on a finite-dimensional manifold, we call the operator Δ_p ($p \geq 0$) the p-th **Laplace–Beltrami operator** associated with the Hilbert complex $(\{\mathscr{H}_p\}_{p=0}^\infty, \{\mathscr{D}_p\}_{p=0}^\infty, \{D_p\}_{p=0}^\infty)$.

Remark 4.2 For $p \geq 1$, the subspace $\mathrm{Dom}(D_p^* D_p) \cap \mathrm{Dom}(D_{p-1} D_{p-1}^*)$ may not be dense in $\mathscr{H}_p$. Hence the operator $L_p := D_p^* D_p + D_{p-1} D_{p-1}^*$ may not be densely defined. Moreover, even if $\mathrm{Dom}(D_p^* D_p) \cap \mathrm{Dom}(D_{p-1} D_{p-1}^*)$ is dense (then L_p turns out to be a symmetric operator), it is non-trivial if L_p is essentially self-adjoint or not. This is the reason why we define the p-th Laplace–Beltrami operator Δ_p via the sesquilinear form q.

Each non-zero element in $\ker \Delta_p$ may be interpreted as an abstract p-th *harmonic form* on $\mathcal{H}_p$.

Lemma 4.5 *For all* $p \geq 0$, $\ker \Delta_p = \hat{\mathcal{H}}_p$. *In particular,* $\dim \ker \Delta_p = \dim \hat{\mathcal{H}}_p$.

Proof We first remark that, for every non-negative self-adjoint operator A on a Hilbert space, $\ker A = \ker A^{1/2}$. Let $\Psi \in \ker \Delta_p$. Then $\Delta_p^{1/2}\Psi = 0$. Hence, by (4.25), $D_p\Psi = 0$ and $D_{p-1}^*\Psi = 0$, implying that $\Psi \in \hat{\mathcal{H}}_p$. Thus $\ker \Delta_p \subset \hat{\mathcal{H}}_p$. Conversely, let $\Psi \in \hat{\mathcal{H}}_p$. Then, by (4.25), $\|\Delta_p^{1/2}\Psi\|^2 = 0$, implying that $\Psi \in \ker \Delta_p^{1/2}$. Hence $\Psi \in \ker \Delta_p$. Thus $\hat{\mathcal{H}}_p \subset \ker \Delta_p$. □

Theorem 4.2 and Lemma 4.5 yield the following result:

Corollary 4.1 (de Rham–Hodge–Kodaira decomposition) *For each* $p \geq 0$, $\mathcal{H}_p$ *has the orthogonal decomposition*

$$\mathcal{H}_p = \ker \Delta_p \oplus \overline{\mathcal{R}_{p-1}} \oplus \overline{\mathcal{R}_p^*}. \tag{4.27}$$

In particular, if $\mathcal{R}_{p-1}$ *and* $\mathcal{R}_p^*$ *are closed, then*

$$\mathcal{H}_p = \ker \Delta_p \oplus \mathcal{R}_{p-1} \oplus \mathcal{R}_p^*. \tag{4.28}$$

Corollary 4.2 *Suppose that, for each* $p \geq 0$, D_p *is semi-Fredholm. Then*

$$\mathcal{H}_p = \ker \Delta_p \oplus \mathcal{R}_{p-1} \oplus \mathcal{R}_p^*.$$

If each D_p *(*$p \geq 0$*) is Fredholm, then* $\dim \ker \Delta_p < \infty$.

Proof Since $\mathrm{Ran}(D_{p-1})$ is closed, we have $\overline{\mathcal{R}_{p-1}} = \mathcal{R}_{p-1}$. It is well-known that, if T is semi-Fredholm (resp. Fredholm), then so is T^* (e.g., [45, Chap. IV, Corollary 5.14]). Hence $\mathrm{Ran}(D_p^*)$ is closed. Therefore $\overline{\mathcal{R}_p^*} = \mathcal{R}_p^*$. Thus (4.27) yields (4.28).

If each D_p ($p \geq 0$) is Fredholm, then $\dim \ker D_p < \infty$ and $\dim \ker D_{p-1}^* < \infty$. Hence $\dim \hat{\mathcal{H}}_p < \infty$. Then, by Lemma 4.5, $\dim \ker \Delta_p < \infty$. □

4.5.3 The Dirac and the Laplace–Beltrami Operators Associated With a Hilbert Complex

Let $(\{\mathcal{H}_p\}_{p=0}^\infty, \{\mathcal{D}_p\}_{p=0}^\infty, \{D_p\}_{p=0}^\infty)$ be a Hilbert complex with Assumption (A). It is natural to construct the infinite direct sum Hilbert space

$$\mathfrak{H} := \oplus_{p=0}^{\infty} \mathscr{H}_p$$

Then there is the right shift operator on $\mathfrak{H}$ [22, p. 165, §4.4] associated with $\{D_p\}_{p=0}^{\infty}$. We denote it by D, which is given as follows:

$$\mathrm{Dom}(D) = \left\{ \Psi \in \mathfrak{H} \,\middle|\, \Psi^{(p)} \in \mathscr{D}_p, \, p \geq 0, \, \sum_{p=1}^{\infty} \|D_{p-1}\Psi^{(p-1)}\|^2 < \infty \right\},$$

$$(D\Psi)^{(p)} = D_{p-1}\Psi^{(p-1)}, \quad p \geq 0, \quad \Psi \in \mathrm{Dom}(D).$$

It follows from a general theorem on right shift operators [22, Theorem 4.3] that D is a densely defined closed linear operator and the adjoint D^* takes the following form:

$$\mathrm{Dom}(D^*) = \left\{ \Psi \in \mathfrak{H} \,\middle|\, \Psi^{(p+1)} \in \mathscr{D}_p^*, \, p \geq 0, \, \sum_{p=0}^{\infty} \|D_p^*\Psi^{(p+1)}\|^2 < \infty \right\},$$

$$(D^*\Psi)^{(p)} := D_p^*\Psi^{(p+1)}, \quad p \geq 0, \, \Psi \in \mathrm{Dom}(D^*).$$

The operators D and D^* are nilpotent in the sense of the next lemma:

Lemma 4.6 $D^2 = 0$ *on* $\mathrm{Dom}(D)$ *and* $(D^*)^2 = 0$ *on* $\mathrm{Dom}(D^*)$.

Proof Let $\Psi \in \mathrm{Dom}(D)$. Then $(D\Psi)^{(0)} = 0$ and $(D\Psi)^{(p)} = D_{p-1}\Psi^{(p-1)}$, $p \geq 1$. Hence, by (4.22), $(D\Psi)^{(p)}$ is in $\mathrm{Dom}(D_p)$ and $D_p(D\Psi)^{(p)} = 0$. This implies that $D\Psi \in \mathrm{Dom}(D)$ and $D(D\Psi) = 0$. Hence $D^2 = 0$ on $\mathrm{Dom}(D)$.

Let $\Phi \in \mathrm{Dom}(D^*)$. Then, by the preceding result, $\langle D^*\Phi, D\Psi \rangle = \langle \Phi, D^2\Psi \rangle = 0 = \langle 0, \Psi \rangle$. Hence $\langle D^*\Phi, D\Psi \rangle = \langle 0, \Psi \rangle$, $\Psi \in \mathrm{Dom}(D)$. This means that $D^*\Phi \in \mathrm{Dom}(D^*)$ and $D^*(D^*\Phi) = 0$. Hence $(D^*)^2 = 0$ on $\mathrm{Dom}(D^*)$. $\quad\square$

In analogy with analysis on finite-dimensional manifolds, it is natural to consider the operator

$$Q := D + D^*. \tag{4.29}$$

We call it the **Dirac operator** associated with D. It follows from Assumption (A) that $\mathrm{Dom}(Q) = \mathrm{Dom}(D) \cap \mathrm{Dom}(D^*)$ is dense in $\mathfrak{H}$. It is obvious that $Q \subset Q^*$. Hence Q is a symmetric operator. But, at this stage, it is unclear if Q is essentially self-adjoint or not. A basic property of Q is:

Lemma 4.7 *The operator* Q *is closed.*

Proof Let $\Psi \in \mathrm{Dom}(Q)$. Then, by Lemma 4.6, we have

$$\|Q\Psi\|^2 = \|D\Psi\|^2 + \|D^*\Psi\|^2. \tag{4.30}$$

This implies the closedness of Q (apply a general theorem [22, Theorem 1.4]). $\quad\square$

The following lemma is often useful to prove the essential self-adjointness of symmetric operators (cf. [56, Chap. 10, Problem 28]):

Lemma 4.8 *Let A be a symmetric operator on a Hilbert space $\mathcal{H}$ such that A^2 is essentially self-adjoint on a dense subspace $\mathcal{D} \subset \mathrm{Dom}(A^2)$. Then A is essentially self-adjoint on $\mathcal{D}$.*

Proof We set $A_{\mathcal{D}} := A \upharpoonright \mathcal{D}$, the restriction of A to $\mathcal{D}$. Since A^2 is non-negative and essentially self-adjoint on $\mathcal{D}$, $\overline{A^2}$ is a non-negative self-adjoint operator. Hence $T := \overline{A^2} + 1$ is a strictly positive self-adjoint operator with $T \geq 1$. By a limiting argument, one can show that $\mathrm{Dom}(T) \subset \mathrm{Dom}(\overline{A}_{\mathcal{D}})$ and, for all $\Psi \in \mathrm{Dom}(T)$, $\langle \overline{A}_{\mathcal{D}} \Psi, T \Psi \rangle = \langle T \Psi, \overline{A}_{\mathcal{D}} \Psi \rangle$. Applying this equation to the case where $\Psi = T^{-1} \Phi$ with $\Phi \in \mathrm{Dom}(A_{\mathcal{D}}^*)$, we have $\langle \Psi, A_{\mathcal{D}}^* \Phi \rangle = \langle A_{\mathcal{D}}^* \Phi, \Psi \rangle$.[1] This implies that, for all $a \in \mathbb{R} \setminus \{0\}$,

$$\mathrm{Im} \langle \Psi, (A_{\mathcal{D}}^* + ia) \Phi \rangle = a \langle \Psi, \Phi \rangle = a \| T^{-1/2} \Phi \|^2.$$

Hence, taking $\Phi \in \ker(A_{\mathcal{D}}^* + ia)$, we have $\| T^{-1/2} \Phi \|^2 = 0$, implying that $T^{-1} \Phi = 0$ and hence $\Phi = 0$. Therefore $\ker(A_{\mathcal{D}}^* + ia) = \{0\}$. In particular, $\ker(A_{\mathcal{D}}^* \pm i) = \{0\}$. Thus, by the well-known criterion (see, e.g., [55, p.257, Corollary]) on essential self-adjointness of symmetric operators, $A_{\mathcal{D}}$ is essentially self-adjoint, i.e., A is essentially self-adjoint on $\mathcal{D}$. $\qquad\square$

The following proposition formulates a sufficient condition for Q to be self-adjoint:

Proposition 4.1 *Suppose that Q^2 is essentially self-adjoint on a dense subspace $\mathfrak{D} \subset \mathrm{Dom}(Q^2)$. Then Q is self-adjoint and essentially self-adjoint on $\mathfrak{D}$.*

Proof By applying Lemma 4.8 to the case $A = Q$, we see that Q is essentially self-adjoint on $\mathfrak{D}$. But, by Lemma 4.7, Q is closed. Hence Q is self-adjoint (cf. [55, Theorem VIII.3, Corollary (p.257)]). $\qquad\square$

The sequence $\{\Delta_p\}_{p=0}^{\infty}$ of the Laplace–Beltrami operators naturally defines an infinite direct sum operator on $\mathcal{H}$:

$$\Delta := \oplus_{p=0}^{\infty} \Delta_p. \tag{4.31}$$

We call it the **Laplace–Beltrami operator** on $\mathfrak{H}$. It follows that Δ is a non-negative self-adjoint operator with

$$\ker \Delta = \oplus_{p=0}^{\infty} \ker \Delta_p.$$

Lemma 4.9 *The following operator equality holds:*

$$\Delta^{1/2} = \oplus_{p=0}^{\infty} \Delta_p^{1/2}.$$

---

[1] Recall that, for any densely defined closable linear operator L, $(\bar{L})^* = L^*$.

Proof This follows from functional calculus of an infinite direct sum operator of self-adjoint operators [22, Theorem 4.4(iii)]. □

Theorem 4.3 *The operator equality*

$$\Delta = Q^*Q \tag{4.32}$$

holds.

Proof In the same way as in the proof of (4.30) (or by (4.30) and the polarization identity), one can show that, for all $\Psi, \Phi \in D(Q)$,

$$\langle Q\Psi, Q\Phi \rangle = \langle D\Psi, D\Phi \rangle + \langle D^*\Psi, D^*\Phi \rangle$$

$$= \sum_{p=0}^{\infty} (\langle D_p \Psi^{(p)}, D_p \Phi^{(p)} \rangle + \langle D_{p-1}^* \Psi^{(p)}, D_{p-1}^* \Phi^{(p)} \rangle)$$

$$= \sum_{p=0}^{\infty} \langle \Delta_p^{1/2} \Psi^{(p)}, \Delta_p^{1/2} \Phi^{(p)} \rangle.$$

By Lemma 4.9, the right-hand side is equal to $\langle \Delta^{1/2}\Psi, \Delta^{1/2}\Phi \rangle$. Hence

$$\langle \Delta^{1/2}\Psi, \Delta^{1/2}\Phi \rangle = \langle Q\Psi, Q\Phi \rangle. \tag{4.33}$$

Note that $D(Q) = D(\Delta^{1/2})$. Hence (4.33) shows that Δ is the self-adjoint operator associated with the sesquilinear form $(\Psi, \Phi) \mapsto \langle Q\Psi, Q\Phi \rangle$ which is closed by Lemma 4.7. Hence (4.32) holds. □

The next corollary immediately follows from (4.32):

Corollary 4.3 *If Q is self-adjoint, then $\Delta = Q^2$.*

4.5.4 Supersymmetric Structure

It is easy to see that the operator

$$\Gamma_{\mathfrak{H}} := \oplus_{p=0}^{\infty}(-1)^p$$

is bounded with $\mathrm{Dom}(\Gamma_{\mathfrak{H}}) = \mathfrak{H}$ satisfying

$$\Gamma_{\mathfrak{H}}^2 = I, \quad \Gamma_{\mathfrak{H}}^* = \Gamma_{\mathfrak{H}}, \quad \Gamma_{\mathfrak{H}} \neq \pm I.$$

In other words, $\Gamma_{\mathfrak{H}}$ is a grading operator on $\mathfrak{H}$.

Lemma 4.10 *For all $\Psi \in \text{Dom}(Q)$, $\Gamma_{\mathfrak{H}}\Psi$ is in $\text{Dom}(Q)$ and*

$$\Gamma_{\mathfrak{H}} Q\Psi + Q\Gamma_{\mathfrak{H}}\Psi = 0. \tag{4.34}$$

Proof Let $\Psi \in \text{Dom}(Q)$. Then $\Gamma_{\mathfrak{H}}\Psi = \{(-1)^p \Psi^{(p)}\}_{p=0}^{\infty}$. Since $|(-1)^p| = 1$, it follows that $\Gamma_{\mathfrak{H}}\Psi \subset \text{Dom}(D) \cap \text{Dom}(D^*) = \text{Dom}(Q)$ and, for all $p \geq 0$,

$$\begin{aligned}
(Q\Gamma_{\mathfrak{H}}\Psi)^{(p)} &= D_{p-1}((-1)^{p-1}\Psi^{(p-1)}) + D_p^*((-1)^{p+1}\Psi^{(p+1)}) \\
&= -(-1)^p (D\Psi)^{(p)} - (-1)^p (D^*\Psi)^{(p)} = -(\Gamma_{\mathfrak{H}} Q\Psi)^{(p)}.
\end{aligned}$$

Hence (4.34) holds. □

Corollary 4.3 and Lemma 4.10 imply:

Theorem 4.4 *Suppose that Q is self-adjoint. Then $(\mathfrak{H}, \Gamma_{\mathfrak{H}}, Q, \Delta)$ is an SQM.*

Thus we see that, for a Hilbert complex such that the Dirac operator Q is self-adjoint, a supersymmetric structure is associated with it.

In Theorem 4.4, the self-adjointness of Q is assumed. But, in the case where Q is not necessarily self-adjoint, we can use the method described in Sect. 1.5.2. The Hilbert space $\mathfrak{H}$ has the following orthogonal decomposition:

$$\mathfrak{H} = \mathfrak{H}_{\text{even}} \oplus \mathfrak{H}_{\text{odd}}$$

with

$$\mathfrak{H}_{\text{even}} := \oplus_{p=0}^{\infty} \mathscr{H}_{2p}, \quad \mathscr{H}_{\text{odd}} := \oplus_{p=0}^{\infty} \mathscr{H}_{2p+1}.$$

Since each of D and D^* can be regarded as elements in $\mathfrak{C}(\mathfrak{H}_{\text{even}}, \mathfrak{H}_{\text{odd}})$ and $\mathfrak{C}(\mathfrak{H}_{\text{odd}}, \mathfrak{H}_{\text{even}})$ (hence so can be Q), we can define $Q_+ \in \mathfrak{C}(\mathscr{H}_{\text{even}}, \mathfrak{H}_{\text{odd}})$ and $Q_- \in \mathfrak{C}(\mathfrak{H}_{\text{odd}}, \mathfrak{H}_{\text{even}})$ by

$$Q_+ := Q \upharpoonright \text{Dom}(Q) \cap \mathfrak{H}_{\text{even}}, \quad Q_- := Q \upharpoonright \text{Dom}(Q) \cap \mathfrak{H}_{\text{odd}},$$

where the closedness of $Q_\pm$ follows from Lemma 4.7. By the results in Sect. 1.5.2, the operators Q_1 and Q_2 defined by

$$Q_1 := \begin{pmatrix} 0 & Q_+^* \\ Q_+ & 0 \end{pmatrix}, \quad Q_2 := \begin{pmatrix} 0 & Q_- \\ Q_-^* & 0 \end{pmatrix}$$

are self-adjoint extensions of Q. It follows that

$$\Delta_1 := Q_1^2 = \begin{pmatrix} Q_+^* Q_+ & 0 \\ 0 & Q_+ Q_+^* \end{pmatrix}$$

and

$$\Delta_2 := Q_2^2 = \begin{pmatrix} Q_- Q_-^* & 0 \\ 0 & Q_-^* Q_- \end{pmatrix}$$

are non-negative self-adjoint operators. It is easy to see that

$$\Gamma_{\mathfrak{H}} := I \oplus (-I)$$

on $\mathfrak{H}$. Then, by the theory in Sect. 1.5.2, we obtain:

Theorem 4.5 *The quadruples* $(\mathfrak{H}, \Gamma_{\mathfrak{H}}, Q_1, \Delta_1)$ *and* $(\mathfrak{H}, \Gamma_{\mathfrak{H}}, Q_2, \Delta_2)$ *are both SQM.*

If Q is self-adjoint, then the two SQM are the same satisfying $Q = Q_1 = Q_2$ and $\Delta = \Delta_1 = \Delta_2$.

4.6 Hilbert Complexes Associated With Boson–Fermion Fock Space

We now come back to the boson–fermion Fock space $\mathscr{F}(\mathscr{H}, \mathscr{K})$. By the theory in Sect. 4.3, we have:

Lemma 4.11 *The triple* $\mathsf{H}_{\mathrm{BF}} := (\{\mathscr{F}^{(p)}(\mathscr{H}, \mathscr{K})\}_{p=0}^{\infty}, \{\mathrm{Dom}(d_{S,p})\}_{p=0}^{\infty}, \{d_{S,p}\}_{p=0}^{\infty})$ *is a Hilbert complex.*

Therefore we can apply the results in Sect. 4.5 to the Hilbert complex H_{BF}. First, Theorem 4.2 yields:

Theorem 4.6 (weak Hodge decomposition) *For each* $p \geq 0$, $\mathscr{F}^{(p)}(\mathscr{H}, \mathscr{K})$ *has the orthogonal decomposition*

$$\mathscr{F}^{(p)}(\mathscr{H}, \mathscr{K}) = \hat{\mathscr{H}}_{S,p} \oplus \overline{\mathrm{Ran}(d_{S,p-1})} \oplus \overline{\mathrm{Ran}(d_{S,p}^*)}$$

with $\hat{\mathscr{H}}_{S,p} := \ker d_{S,p} \cap \ker d_{S,p-1}^*$.

Lemma 4.12 *For each* $p \geq 0$, $D(d_{S,p}) \cap D(d_{S,p-1}^*)$ *is dense in* $\mathscr{F}^{(p)}(\mathscr{H}, \mathscr{K})$.

Proof By Lemmas 4.1 and 4.2, we have

$$\mathscr{F}_{\mathrm{b,fin}}(\mathrm{Dom}(S)) \hat{\otimes} \wedge^p (\mathrm{Dom}(S^*)) \subset D(d_{S,p}) \cap D(d_{S,p-1}^*).$$

Since $\mathrm{Dom}(S)$ and $\mathrm{Dom}(S^*)$ are dense in $\mathscr{H}$ and $\mathscr{K}$ respectively, the subspace $\mathscr{F}_{\mathrm{b,fin}}(\mathrm{Dom}(S)) \hat{\otimes} \wedge^p (\mathrm{Dom}(S^*))$ is dense in $\mathscr{F}^{(p)}(\mathscr{H}, \mathscr{K})$. Thus the desired result follows. $\qquad \square$

By Lemma 4.12, Assumption (A) in Sect. 4.5 is satisfied in the present context $(\mathscr{D}_p = \mathrm{Dom}(d_{S,p}), \mathscr{D}_{p-1}^* = \mathrm{Dom}(d_{S,p-1}^*))$. Hence, by Lemma 4.4, there exists a unique non-negative self-adjoint operator $\Delta_{S,p}$ on $\mathscr{F}^{(p)}(\mathscr{H}, \mathscr{K})$ such that

$$\mathrm{Dom}(\Delta_{S,p}^{1/2}) = D(d_{S,p}) \cap D(d_{S,p-1}^*)$$

and, for all $\Psi, \Phi \in D(d_{S,p}) \cap D(d_{S,p-1}^*)$,

$$\left\langle \Delta_{S,p}^{1/2}\Psi, \Delta_{S,p}^{1/2}\Phi \right\rangle = \left\langle d_{S,p}\Psi, d_{S,p}\Phi \right\rangle + \left\langle d_{S,p-1}^*\Psi, d_{S,p-1}^*\Phi \right\rangle.$$

We call $\Delta_{S,p}$ the p-th **Laplace–Beltrami operator** associated with $S \in \mathfrak{C}(\mathscr{H}, \mathscr{K})$ on $\mathscr{F}^{(p)}(\mathscr{H}, \mathscr{K})$. In particular, $\Delta_{S,0}$ is called the **Laplacian** on $\mathscr{F}^{(0)}(\mathscr{H}, \mathscr{K}) \cong \mathscr{F}_{\mathrm{b}}(\mathscr{H})$ associated with S.

We next want to identify $\Delta_{S,p}$ in terms of known operators. The subspace

$$\mathscr{D}_S^{(p)} := \mathscr{F}_{\mathrm{b,fin}}(\mathrm{Dom}(S^*S)) \hat{\otimes} \wedge^p(\mathrm{Dom}(SS^*)) \tag{4.35}$$

is dense in $\mathscr{F}^{(p)}(\mathscr{H}, \mathscr{K})$ and included in $\mathrm{Dom}(d_{S,p}^* d_{S,p}) \cap \mathrm{Dom}(d_{S,p-1} d_{S,p-1}^*)$ (see Lemmas 4.1 and 4.2). Hence it follows that the operator

$$L_{S,p} := d_{S,p}^* d_{S,p} + d_{S,p-1} d_{S,p-1}^*.$$

is a non-negative symmetric operator on $\mathscr{F}^{(p)}(\mathscr{H}, \mathscr{K})$ $(d_{S,-1} := 0)$.

Lemma 4.13 *For all $p \geq 0$, $L_{S,p}$ is closed.*

Proof For all $\Psi \in \mathscr{D}_S^{(p)}$, we have by (4.19)

$$\|L_{S,p}\Psi\|^2 = \|d_{S,p}^* d_{S,p}\Psi\|^2 + \|d_{S,p-1} d_{S,p-1}^*\Psi\|^2.$$

Since $d_{S,p}^* d_{S,p}$ and $d_{S,p-1} d_{S,p-1}^*$ are non-negative self-adjoint and hence closed, the equation implies that $L_{S,p}$ is closed. $\qquad\square$

Theorem 4.7 *For all $p \geq 0$, $L_{S,p}$ is self-adjoint and the following operator equalities hold:*

$$\Delta_{S,p} = L_{S,p} = d\Gamma_{\mathrm{b}}(S^*S) \otimes I + I \otimes d\Gamma_{\mathrm{f}}^{(p)}(SS^*). \tag{4.36}$$

Proof Let

$$G_p := d\Gamma_{\mathrm{b}}(S^*S) \otimes I + I \otimes d\Gamma_{\mathrm{f}}^{(p)}(SS^*)).$$

Then G_p is non-negative self-adjoint. By Lemma 4.3(iii), we have $L_{S,p} = G_p$ on $\mathscr{D}_S^{(p)}$. Hence $G_p \restriction \mathscr{D}_S^{(p)} \subset L_{S,p}$. On the other hand, $\mathscr{D}_S^{(p)}$ is a core for G_p [22, Theorem 4.7 (iii), Theorem 6.1 (iv)]. Hence $G_p \subset \bar{L}_{S,p} = L_{S,p}$ (by Lemma 4.13). Since every self-adjoint operator has no non-trivial symmetric extension, it follows that $G_p = L_{S,p}$. Thus $L_{S,p}$ is self-adjoint and the second equality in (4.36) holds. By (4.26), we have $L_{S,p} \subset \Delta_{S,p}$. Since $L_{S,p}$ is self-adjoint as already seen. it follows that $L_{S,p} = \Delta_{S,p}$.   □

It follows from (4.36) and a general spectral theory on tensor products of self-adjoint operators [22, Theorem 3.8] that

$$\sigma(\Delta_{S,p}) = \overline{\{\lambda + \mu | \lambda \in \sigma(d\Gamma_b(S^*S)), \mu \in \sigma(d\Gamma_f^{(p)}(SS^*))\}},$$

$$\sigma_p(\Delta_{S,p}) = \{\lambda + \mu | \lambda \in \sigma_p(d\Gamma_b(S^*S)), \mu \in \sigma_p(d\Gamma_f^{(p)}(SS^*))\}.$$

The spectral properties of $d\Gamma_b(\cdot)$ and $d\Gamma_f(\cdot)$ are well-known [22, Theorem 5.3, §6.5]. Hence the spectrum and the point spectrum of $\Delta_{S,p}$ are computed, although we do not write them down here.

In the present context, Corollaries 4.1 and 4.2 take the following forms:

Corollary 4.4 (de Rham–Hodge–Kodaira decomposition) *For each $p \geq 0$, the Hilbert space $\mathscr{F}^{(p)}(\mathscr{H}, \mathscr{K})$ has the orthogonal decomposition*

$$\mathscr{F}^{(p)}(\mathscr{H}, \mathscr{K}) = \ker \Delta_{S,p} \oplus \overline{\mathrm{Ran}(d_{S,p-1})} \oplus \overline{\mathrm{Ran}(d_{S,p}^*)}.$$

In particular, if $\mathrm{Ran}(d_{S,p-1})$ and $\mathrm{Ran}(d_{S,p}^*)$ are closed, then

$$\mathscr{F}^{(p)}(\mathscr{H}, \mathscr{K}) = \ker \Delta_{S,p} \oplus \mathrm{Ran}(d_{S,p-1}) \oplus \mathrm{Ran}(d_{S,p}^*).$$

Corollary 4.5 *Suppose that, for each $p \geq 0$, $d_{S,p}$ is semi-Fredholm. Then*

$$\mathscr{F}^{(p)}(\mathscr{H}, \mathscr{K}) = \ker \Delta_{S,p} \oplus \mathrm{Ran}(d_{S,p-1}) \oplus \mathrm{Ran}(d_{S,p}^*).$$

If each $d_{S,p}$ ($p \geq 0$) is Fredholm, then $\dim \ker \Delta_{S,p} < \infty$.

One can identify $\ker \Delta_{S,p}$:

Theorem 4.8

$$\ker \Delta_{S,p} = \oplus_{n=0}^{\infty} \{(\otimes_s^n \ker S) \otimes \wedge^p(\ker S^*)\}. \tag{4.37}$$

Proof By (4.36) and a general theorem,[2] we have

$$\ker \Delta_{S,p} = \ker d\Gamma_b(S^*S) \otimes \ker d\Gamma_f^{(p)}(SS^*).$$

It is easy to see that

---

[2] Let A_i ($i = 1, 2$) be a non-negative self-adjoint operator on a Hilbert space $\mathscr{H}_i$. Then $\ker(A_1 \otimes I + I \otimes A_2) = \ker(A_1 \otimes A_2)$. See, e.g., [22, Theorem 7.6].

$$\ker d\Gamma_{\mathrm{b}}(S^*S) = \oplus_{n=0}^{\infty} \ker(S^*S)_{\mathrm{s}}^{(n)}.$$

Moreover,

$$\ker(S^*S)_{\mathrm{s}}^{(n)} = \otimes_{\mathrm{s}}^n \ker S^*S = \otimes_{\mathrm{s}}^n \ker S$$

and

$$\ker d\Gamma_{\mathrm{f}}^{(p)}(SS^*) = \wedge^p(\ker(SS^*)) = \wedge^p(\ker S^*).$$

Thus (4.37) holds.   □

For a linear operator T on a Hilbert space, the number

$$\mathrm{nul}\,(T) := \dim \ker T \in \mathbb{Z}_+ \cup \{\infty\}$$

is called the nullity of T.

Formula (4.37) implies:

Corollary 4.6

(i) *If* $\mathrm{nul}\,(S) = 0$, *then*

$$\mathrm{nul}\,(\Delta_{S,p}) = \begin{cases} 1 & (p=0) \\[2mm] \dfrac{\mathrm{nul}\,(S^*)!}{(\mathrm{nul}\,(S^*) - p)!\,p!} & (1 \le p \le \mathrm{nul}\,(S^*) < \infty) \\[2mm] \infty & (\mathrm{nul}\,(S^*) = \infty) \\[2mm] 0 & (0 \le \mathrm{nul}\,(S^*) < p) \end{cases}.$$

(ii) *If* $\mathrm{nul}\,(S) \ge 1$, *then*

$$\mathrm{nul}\,(\Delta_{S,p}) = \begin{cases} \infty & (0 \le p \le \mathrm{nul}\,(S^*) \le \infty) \\[2mm] 0 & (0 \le \mathrm{nul}\,(S^*) < p) \end{cases}.$$

Remark 4.3 (i) The operators $d_{S,p}$ and $\Delta_{S,p}$ were first introduced in [4] in the case where $\mathscr{H} = \{f \in \mathscr{S}'(\mathbb{R}^n) | (-\Delta_n + m^2)^{-1/2} f \in L^2(\mathbb{R}^n)\}$ ($\mathscr{S}'(\mathbb{R}^n)$ is the space of tempered distributions on $\mathbb{R}^n$, Δ_n is the n-dimensional Laplacian and $m > 0$ is a constant), $\mathscr{K} = L^2(\mathbb{R}^n)$ and $S = (-\Delta_n + m^2)^{1/2}$. The first form was generalized in [6, 8, 9]. In [60], the case where $\mathscr{H} = \mathscr{K}$ and $S = I$ was studied from viewpoints of the Malliavin calculus. Fundamental (Sobolev type) spaces of differential forms based on $\Delta_{S,p}$ were proposed and analyzed in [25, 26].

(ii) L. Gross [35] introduced an infinite-dimensional Laplacian in the framework of the abstract Wiener space, which, in the present framework, corresponds to $\Delta_{S,0}$ with $\mathscr{H} = \mathscr{K}$ and $S = I$ and a generalization of it (see also [59, 60]).

4.7 Laplace–Beltrami Operators on Boson–Fermion Fock Space

We denote by Δ_S the Laplace–Beltrami operator on $\mathscr{F}(\mathscr{H}, \mathscr{K})$ (see (4.31)):

$$\Delta_S := \oplus_{p=0}^{\infty} \Delta_{S,p}.$$

By (4.36), we have

$$\Delta_S = d\Gamma_{\mathrm{b}}(S^*S) \otimes I + I \otimes d\Gamma_{\mathrm{f}}(SS^*). \tag{4.38}$$

This may be a remarkable formula, clarifying a "geometric" meaning of the sum of a boson second quantization operator and a fermion second quantization $d\Gamma_{\mathrm{b}}(S^*S) \otimes I + I \otimes d\Gamma_{\mathrm{f}}(SS^*)$.

The sequence $\{\mathscr{D}_S^{(p)}\}_{p=0}^{\infty}$ (see (4.35)) yields the algebraic infinite direct sum

$$\mathscr{D}_S := \hat{\oplus}_{p=0}^{\infty} \mathscr{D}_S^{(p)}. \tag{4.39}$$

It is easy to see that

$$\mathscr{D}_S = \mathscr{F}_{\mathrm{b,fin}}(\mathrm{Dom}(S^*S)) \hat{\otimes} \mathscr{F}_{\mathrm{f,fin}}(\mathrm{Dom}(SS^*))$$

Lemma 4.14 *Let A_1 and A_2 be self-adjoint operators on Hilbert spaces $\mathscr{H}_1$ and $\mathscr{H}_2$ respectively. Suppose that A_i ($i = 1, 2$) is essentially self-adjoint on a dense subspace $\mathscr{D}_i \subset \mathrm{Dom}(A_i)$. Then $A_1 \otimes I + I \otimes A_2$ is essentially self-adjoint on $\mathscr{D}_1 \hat{\otimes} \mathscr{D}_2$.*

Proof See [55, p. 301, Corollary (a)] or [22, Theorem 3.8(iii)]. $\square$

Theorem 4.9 *The Laplace–Beltrami operator Δ_S is essentially self-adjoint on $\mathscr{D}_S$.*

Proof The operator $d\Gamma_{\mathrm{b}}(S^*S)$ is essentially self-adjoint on $\mathscr{F}_{\mathrm{b,fin}}(\mathrm{Dom}(S^*S))$. Similarly, $d\Gamma_{\mathrm{f}}(SS^*)$ is essentially self-adjoint on $\mathscr{F}_{\mathrm{f,fin}}(\mathrm{Dom}(SS^*))$. Hence, by (4.38) and Lemma 4.14, Δ_S is essentially self-adjoint on $\mathscr{D}_S$. $\square$

4.8 Dirac Operators on Boson–Fermion Fock Space

We denote by Q_S the Dirac operator associated with d_S (see (4.29)):

$$Q_S := d_S + d_S^*.$$

An important fact is:

Theorem 4.10 *The operator Q_S is self-adjoint and*

$$\Delta_S = Q_S^2. \tag{4.40}$$

Moreover, Q_S is essentially self-adjoint on $\mathcal{D}_S$.

Proof It is easy to see that $\mathcal{D}_S \subset D(Q_S^2)$. A simple application of (4.32) yields that $\Delta_S = Q_S^2$ on $\mathcal{D}_S$. By this fact and Theorem 4.9, Q_S^2 is essentially self-adjoint on $\mathcal{D}_S$. By this result and Proposition 4.1, Q_S is self-adjoint and essentially self-adjoint on $\mathcal{D}_S$. Then (4.32) implies (4.40). □

Theorems 4.10 and 4.4 imply:

Theorem 4.11 *For each $S \in \mathfrak{C}(\mathcal{H}, \mathcal{K})$, $(\mathcal{F}(\mathcal{H}, \mathcal{K}), \Gamma_{bf}, Q_S, \Delta_S)$ is an SQM.*

By (4.40) and (4.38), we have

$$Q_S^2 = d\Gamma_b(S^*S) \otimes I + I \otimes d\Gamma_f(SS^*). \tag{4.41}$$

This formula gives a basic relation between the Dirac operator Q_S and the second quantization operators $d\Gamma_b(S^*S)$ and $d\Gamma_f(SS^*)$.

It follows from (4.41) and (2.4) that $\Omega_\mathcal{H} \otimes \Omega_\mathcal{K} \in \ker Q_S$. Hence, in the SQM $(\mathcal{F}(\mathcal{H}, \mathcal{K}), \Gamma_{bf}, Q_S, \Delta_S)$, supersymmetry is not spontaneously broken. As is shown later in Chap. 5, the SQM $(\mathcal{F}(\mathcal{H}, \mathcal{K}), \Gamma_{bf}, Q_S, \Delta_S)$ is an abstract form of some free supersymmetric quantum field models. By (1.17) applied to the case $A = S$, we have $\sigma(S^*S) \setminus \{0\} = \sigma(SS^*) \setminus \{0\}$. Hence, in the present abstract model, the "boson mass" $\inf \sigma(S^*S) \setminus \{0\}$ coincides with the "fermion mass" $\inf \sigma(SS^*) \setminus \{0\}$. This is a characteristic of free relativistic supersymmetric quantum field models.

Remark 4.4 By (4.38) and the spectral theory of tensor products of self-adjoint operators [22, Theorem 3.8 (i), Theorem 3.12 (i)], we have

$$\sigma(\Delta_S) = \overline{\{\lambda_b + \lambda_f \,|\, \lambda_b \in \sigma(d\Gamma_b(S^*S)), \lambda_f \in \sigma(d\Gamma_f(SS^*))\}},$$
$$\sigma_p(\Delta_S) = \{\lambda_b + \lambda_f \,|\, \lambda_b \in \sigma_p(d\Gamma_b(S^*S)), \lambda_f \in \sigma_p(d\Gamma_f(SS^*))\}.$$

Hence the spectra of Δ_S are determined by the spectral properties of S^*S and SS^* (for spectral properties of second quantization operators, see [22, Theorem 5.3, Theorem 6.10]). By Theorem 1.2, we have

$$\sigma(Q_S) = \{\pm\sqrt{\mu} \,|\, \mu \in \sigma(\Delta_S)\}, \quad \sigma_p(Q_S) = \{\pm\sqrt{\mu} \,|\, \mu \in \sigma_p(\Delta_S)\}.$$

Thus one can know the spectra of Q_S from those of Δ_S, although we do not write them down here (see [21]).

Remark 4.5 The theory of infinite-dimensional Dirac operators presented here can be extended to study the geometry of non-flat infinite-dimensional manifolds [48–50].

4.9 Strong Anti-Commutativity of Dirac Operators

We have a family $\{Q_S\}_{S \in \mathfrak{C}(\mathcal{H}, \mathcal{K})}$ of Dirac operators indexed by $\mathfrak{C}(\mathcal{H}, \mathcal{K})$. It may be interesting to find properties of this family. From the viewpoint of SQM, it would be natural to ask when Q_S anti-commutes with Q_T ($S, T \in \mathfrak{C}(\mathcal{H}, \mathcal{K})$).

Let

$$\mathcal{D}_{S,T} := \mathscr{F}_{\text{fin}}(\text{Dom}(T^*S) \cap \text{Dom}(S^*T), \text{Dom}(ST^*) \cap \text{Dom}(TS^*)).$$

Then, by Lemma 4.3, $\mathcal{D}_{S,T} \subset \text{Dom}(Q_S Q_T) \cap \text{Dom}(Q_T Q_S)$ and

$$\{Q_S, Q_T\} = d\Gamma_{\text{b}}^{(\text{alg})}(T^*S + S^*T) \hat{\otimes} I + I \hat{\otimes} d\Gamma_{\text{f}}^{(\text{alg})}(ST^* + TS^*) \quad \text{on } \mathcal{D}_{S,T}.$$

This implies:

Lemma 4.15 *The operators Q_S and Q_T anti-commute on $\mathcal{D}_{S,T}$ if and only if*

$$T^*S + S^*T = 0, \quad ST^* + TS^* = 0. \tag{4.42}$$

For self-adjoint operators, there is a concept of anti-commutativity stronger than the usual one, which was introduced by Vasilescu [65]:

Definition 4.1 Self-adjoint operators A and B on a Hilbert space $\mathcal{L}$ are said to **strongly anti-commute** if, for all $t \in \mathbb{R}$, $e^{itB}A \subset Ae^{-itB}$.

Remark 4.6 (i) It is shown a posteriori [65] that this definition is symmetric with respect to A and B so that it is certainly meaningful.
(ii) It is shown also that strongly anti-commuting self-adjoint operators are anti-commuting in the usual sense. But the converse is not true.
(iii) One can prove that self-adjoint operators A and B strongly anti-commute if and only if the operator equality $e^{itB}Ae^{itB} = A$ holds for all $t \in \mathbb{R}$.

It is known that the concept of strong anti-commutativity is useful [13, 14, 65]. Hence it may be important to know when Q_S and Q_T strongly anti-commute. To state a result on this problem, we introduce a self-adjoint operator

$$\Lambda_S := \begin{pmatrix} 0 & S^* \\ S & 0 \end{pmatrix}$$

on the direct sum Hilbert space $\mathcal{H} \oplus \mathcal{K}$. Note that (4.42) is equivalent to

$$\Lambda_S \Lambda_T + \Lambda_T \Lambda_S = 0.$$

Concerning strong anti-commutativity of Dirac operators $\{Q_S\}_{S \in \mathfrak{C}(\mathcal{H}, \mathcal{K})}$, there is a beautiful structure:

Theorem 4.12 *The Dirac operators Q_S and Q_T strongly anti-commute if and only if Λ_S and Λ_T strongly anti-commute. In that case, $S \pm T \in \mathfrak{C}(\mathscr{H}, \mathscr{K})$ and $Q_{S \pm T} = Q_S \pm Q_T$.*

Proof See [16, Theorem 3.1]. □

Theorem 4.12 can be applied to construction of integrable representations, on the boson–fermion Fock space, of the two-dimensional relativistic supersymmetry algebra which is generated by four elements Q_1, Q_2, H, P with defining relations

$$Q_1^2 = H + P, \quad Q_2^2 = H - P, \quad Q_1 Q_2 + Q_2 Q_1 = 0.$$

For the details, see [13, 16].

Remark 4.7 The following hold [16, Lemma 3.4]: (i) Q_S strongly commutes with the total number operator N_{tot} (see (4.4)); (ii) Q_S is reduced by each $\mathscr{F}_r(\mathscr{H}, \mathscr{K}), r \in \mathbb{Z}_+$ (see (4.2)); (iii) the operator Λ_S is unitarily equivalent to $Q_{S,1}$, the reduced part of Q_S to $\mathscr{F}_1(\mathscr{H}, \mathscr{K})$, in a natural way. These properties of Q_S also are interesting.

4.10 Perturbations of Dirac Operator Q_S

A method to construct an interacting supersymmetric quantum field model in the present abstract framework is to define a new Dirac operator as a perturbation of Q_S by a suitable operator. A natural way to define a perturbation of Q_S is to perturb d_S by a suitable operator. For this purpose, it is more convenient to work with the Q-space representation $\mathfrak{F}$ of the boson–fermion Fock space $\mathscr{F}(\mathscr{H}, \mathscr{K})$ (see (4.6)). As is already remarked, each linear operator L on $\mathscr{F}(\mathscr{H}, \mathscr{K})$ has the form $V_{bf} L V_{bf}^{-1}$ in the Q-space representation. But, in what follows, for notational simplicity, we use the symbol L for $V_{bf} L V_{bf}^{-1}$ also unless otherwise stated.

4.10.1 Witten Deformation

Let $W : Q_{\mathfrak{h}} \mapsto \mathbb{R} \cup \{\pm\infty\}$ be measurable and finite a.e.. Then a perturbation of d_S is given by

$$d_{S,W} := e^{-W} d_S e^{W},$$

where $e^{\pm W}$ are regarded as multiplication operators on $\mathfrak{F}$. This is called the **Witten deformation** of d_S.[3] It is easy to see that, if W is bounded, then $d_{S,W}$ is densely defined

---

[3] This deformation was introduced by Witten [69] for the exterior differential operator on a finite-dimensional manifold.

and closed. But, in the case where W is unbounded, it may depend on the properties of W whether or not $d_{S,W}$ is densely defined. We do not discuss this problem here; we simply *assume that $d_{S,W}$ is densely defined and closable*. We denote its closure by the same symbol. Then the adjoint $d_{S,W}^*$ exists and is densely defined, satisfying $d_{S,W}^* \supset e^W d_S^* e^{-W}$. It follows that

$$d_{S,W}^2 = 0 \quad \text{on Dom}(d_{S,W}).$$

A perturbation of Q_S is defined by

$$Q_{S,W} := d_{S,W} + d_{S,W}^*.$$

It follows that $Q_{S,W}$ is closed. If $\text{Dom}(d_{S,W}) \cap \text{Dom}(d_{S,W}^*)$ is dense, then $Q_{S,W}$ is a closed symmetric operator.

Suppose that $\text{Dom}(d_{S,W}) \cap \text{Dom}(d_{S,W}^*)$ is dense. Then, as in Lemma 4.4, there exists a unique non-negative self-adjoint operator $\Delta_{S,W}$ on $\mathfrak{F}$ such that $\text{Dom}(\Delta_{S,w}^{1/2}) = \text{Dom}(d_{S,W}) \cap \text{Dom}(d_{S,W}^*)$ and, for all $\Psi, \Phi \in \text{Dom}(d_{S,W}) \cap \text{Dom}(d_{S,W}^*)$,

$$\left\langle \Delta_{S,w}^{1/2} \Psi, \Delta_{S,W}^{1/2} \Phi \right\rangle = \left\langle d_{S,W} \Psi, d_{S,W} \Phi \right\rangle + \left\langle d_{S,W}^* \Psi, d_{S,W}^* \Phi \right\rangle.$$

The operator $\Delta_{S,W}$ gives a perturbation of the Laplace–Beltrami operator Δ_S.

Using (4.21), one has formally

$$d_{S,W} = d_S + \frac{1}{\sqrt{2}} B(S \nabla W(q))^* \quad \text{(a.e. } q \in Q_\mathfrak{h}), \tag{4.43}$$

where $B(\cdot)^*$ is the fermion creation operator on $\mathscr{F}_f(\mathscr{K})$ and ∇ is the gradient operator on $L^2(Q_\mathfrak{h}, d\mu_\mathfrak{h})$ (see (3.6) and (3.7)). In this form, one can extend the perturbation of d_S to a more general one by replacing the $\mathscr{K}$-valued function $S \nabla W / \sqrt{2}$ on $Q_\mathfrak{h}$ with a general one $F : Q_\mathfrak{h} \to \mathscr{K}$. In the next subsection, we consider this type of perturbation.

Remark 4.8 The perturbation of d_S given above is formally regarded as a special case of the more general perturbation which replaces the measure $\mu_\mathfrak{h}$ with a general one [8]. This point of view was further developed in [1–3]. In the framework of an abstract Wiener space, spectral analysis of an operator corresponding to the Laplace–Beltrami operator $\Delta_{I,W}$ (the case where $\mathscr{K} = \mathscr{H}$ and $S = I$) is made in [37] with W obeying a set of conditions.

Remark 4.9 There is another type of perturbation for d_S: $d_S(\alpha) := d_S + \alpha A(g) \otimes B(v)^*$ with $\alpha \in \mathbb{C}$, $g \in \mathscr{H}$, $v \in \mathscr{K}$. This model is explicitly analyzable and one can see that the model has some interesting features. See [21] for the details.

4.10.2 More General Perturbations

Let $F \in L^2(Q_\mathfrak{h}, d\mu_\mathfrak{h}; \mathscr{K})$ and $B(u)$ $(u \in \mathscr{K})$ be the fermion annihilation operator on $\mathscr{F}_f(\mathscr{K})$ with test vector u (see Sect. 2.11). Then, for a.e. q, $F(q)$ is in $\mathscr{K}$ and $\int_{Q_\mathfrak{h}} \|F(q)\|_{\mathscr{K}}^2 d\mu_\mathfrak{h}(q) < \infty$. Hence $B(F(q))$ is a bounded linear operator on $\mathscr{F}_f(\mathscr{K})$ and, by (2.21)

$$\int_{Q_\mathfrak{h}} \|B(F(q))^\#\|^2 d\mu_\mathfrak{h}(q) = \|F\|^2, \tag{4.44}$$

where $B(\cdot)^\#$ denotes either $B(\cdot)$ or $B(\cdot)^*$. We define an operator $B(F)$ on $\mathfrak{F}$ as follows:

$$\mathrm{Dom}(B(F)) := \left\{ \Psi \in \mathfrak{F} \mid \int_{Q_\mathfrak{h}} \|B(F(q))\Psi(q)\|^2 d\mu_\mathfrak{h}(q) < \infty \right\},$$

$$(B(F)\Psi)(q) := B(F(q))\Psi(q), \quad \Psi \in \mathfrak{F}, \text{ a.e. } q \in Q_\mathfrak{h}.$$

Note that $B(F)$ is not necessarily bounded.[4]

Let $M_{\|F\|}$ be the multiplication operator on $L^2(Q_\mathfrak{h}, d\mu_\mathfrak{h})$ by the function $\|F(\cdot)\|$. Then $\mathrm{Dom}(M_{\|F\|})$ is dense in $L^2(Q_\mathfrak{h}, d\mu_\mathfrak{h})$. It follows from (2.21) that

$$\mathrm{Dom}(M_{\|F\|}) \hat{\otimes} \mathscr{F}_f(\mathscr{K}) \subset \mathrm{Dom}(B(F)).$$

Hence $\mathrm{Dom}(B(F))$ is dense in $\mathfrak{F}$. Therefore the adjoint $B(F)^*$ exists. It is not so difficult to show that

$$\mathrm{Dom}(B(F)^*) = \left\{ \Psi \in \mathfrak{F} \mid \int_{Q_\mathfrak{h}} \|(B(F(q))^*\Psi(q)\|^2 d\mu_\mathfrak{h}(q) < \infty \right\},$$

$$(B(F)^*\Psi)(q) = B(F(q))^*\Psi(q), \quad \Psi \in \mathrm{Dom}(B(F)^*), \text{ a.e. } q \in Q_\mathfrak{h}.$$

We now consider the following perturbation of d_S:

$$d_S(F) := d_S + B(F)^*.$$

In the Q-space representation $\mathfrak{F}$ of $\mathscr{F}(\mathscr{H}, \mathscr{K})$, we have by (4.21)

---

[4] A simple example is given by the case where $F(q) = f(q)u$ for an a.e. finite function f on $Q_\mathfrak{h}$ and $u \in \mathscr{K}$. In this case, $B(F(q)) = f(q)^* B(u)$. If f is not essentially bounded on $(Q_\mathfrak{h}, \mu_\mathfrak{h})$, then the multiplication operator by f^* is unbounded and hence $B(F)$ is unbounded.

$$d_S = \frac{1}{\sqrt{2}} \sum_{n=1}^{\infty} B(\xi_n)^* D_{S^*\xi_n} \quad \text{on } \mathscr{P}(\text{Dom}(S)) \hat{\otimes} \mathscr{F}_{f,\text{fin}}(\mathscr{K}). \tag{4.45}$$

In what follows, we assume the following:

(F.1) $U_b \mathscr{F}_{b,\text{fin}}(\mathscr{H}) \subset \text{Dom}(M_{\|F\|})$.

It follows from Assumption (F.1) that

$$\text{Dom}(d_S(F)) \supset V_{bf}(\mathscr{F}_{b,\text{fin}}(\text{Dom}(S)) \hat{\otimes} \mathscr{F}_{f,\text{fin}}(\mathscr{K})).$$

Hence $d_S(F)$ is densely defined. Therefore the adjoint $d_S(F)^*$ exists and satisfies

$$d_S(F)^* \supset d_S^* + B(F).$$

This implies that $d_S(F)^*$ also is densely defined. Hence $d_S(F)$ is closable. With this preliminary, we define a perturbation of Q_S by

$$Q_S(F) := d_S(F) + d_S(F)^*. \tag{4.46}$$

Let

$$\hat{\mathscr{D}}_S := V_{bf} \mathscr{D}_S,$$

where $\mathscr{D}_S$ is defined by (4.39). Then $\hat{\mathscr{D}}_S \subset \text{Dom}(Q_S(F))$ and

$$Q_S(F) = Q_S + B(F) + B(F)^* \quad \text{on } \hat{\mathscr{D}}_S. \tag{4.47}$$

Hence $Q_S(F)$ is densely defined and a symmetric operator on $\mathfrak{F}$. Equation (4.47) shows that $Q_S(F) \upharpoonright \hat{\mathscr{D}}_S$ is a perturbation of Q_S by the operator $B(F) + B(F)^*$. As we have seen in Theorem 4.10, Q_S is essentially self-adjoint on $\hat{\mathscr{D}}_S$. But it may depend on properties of F whether or not $Q_S(F)$ is essentially self-adjoint on a suitable subspace. A simple case is given in the following proposition:

Proposition 4.2 *Suppose that the function* $\|F(\cdot)\|_{\mathscr{K}}$ *on* $Q_{\mathfrak{h}}$ *is essentially bounded. Then* $Q_S(F)$ *is self-adjoint with* $\text{Dom}(Q_S(F)) = \text{Dom}(Q_S)$ *and essentially self-adjoint on any core of* Q_S.

Proof Under the present assumption, $B(F)$ and $B(F)^*$ are bounded and so is $B(F) + B(F)^*$. Hence the assertions follow from the Kato–Rellich theorem. □

In the case where the assumption in Proposition 4.2 does not hold, however, it becomes a highly mathematical problem to prove essential self-adjointness of $Q_S(F)$. A partial result on this problem is obtained in [12].

For notational simplicity, *we denote the closure of* $Q_S(F) \upharpoonright \hat{\mathscr{D}}_S$ *by the same symbol* $Q_S(F)$. Under Assumptions (F.1), we have the following:

Proposition 4.3 *For all $\Psi \in \mathrm{Dom}(Q_S(F))$, $\Gamma_{bf}\Psi$ is in $\mathrm{Dom}(Q_S(F))$ and $\Gamma_{bf}Q_S(F)\Psi = -Q_S(F)\Gamma_{bf}\Psi$.*

Proof It is easy to show that $\Gamma_{bf}Q_S(F)\Psi = -Q_S(F)\Gamma_{bf}\Psi$, $\Psi \in \hat{\mathscr{D}}_S$. Then the statement is proved by a simple limiting argument.   □

Proposition 4.3 implies that $(\mathfrak{F}, \Gamma_{bf}, Q_S(F), H_S(F))$, where

$$H_S(F) := Q_S(F)^2,$$

is an SQM if $Q_S(F)$ *is self-adjoint*. This SQM is regarded as an abstract supersymmetric quantum field model whose supercharge and supersymmetric Hamiltonian are given by $Q_S(F)$ and $H_S(F)$ respectively.

The orthogonal decomposition (2.2) of a fermion Fock space induces the orthogonal decomposition

$$\mathfrak{F} = \mathfrak{F}_+ \oplus \mathfrak{F}_-, \tag{4.48}$$

where

$$\mathfrak{F}_+ := L^2(Q_{\mathfrak{h}}, d\mu_{\mathfrak{h}}; \mathscr{F}_{f,+}(\mathscr{K})), \quad \mathfrak{F}_- := L^2(Q_{\mathfrak{h}}, d\mu_{\mathfrak{h}}; \mathscr{F}_{f,-}(\mathscr{K})).$$

Then we have

$$\Gamma_{bf} \restriction \mathfrak{F}_\pm = \pm I$$

and $Q_S(F)$ has the operator matrix representation with respect to (4.48)

$$Q_S(F) = \begin{pmatrix} 0 & Q_{S,-}(F) \\ Q_{S,+}(F) & 0 \end{pmatrix},$$

where $Q_{S,\pm}(F) := Q_S(F) \restriction \mathrm{Dom}(Q_S(F)) \cap \mathfrak{F}_\pm$. Following the prescription in Sect. 1.5.2, one can define two self-adjoint operators:

$$Q_S^{(1)}(F) := \begin{pmatrix} 0 & Q_{S,+}(F)^* \\ Q_{S,+}(F) & 0 \end{pmatrix}, \quad Q_S^{(2)}(F) := \begin{pmatrix} 0 & Q_{S,-}(F) \\ Q_{S,-}^*(F) & 0 \end{pmatrix}.$$

Then $Q_S^{(1)}(F)$ and $Q_S^{(2)}(F)$ are self-adjoint extensions of $Q_S(F)$. Hence, letting

$$H_S^{(1)}(F) := Q_S^{(1)}(F)^2, \quad H_S^{(2)}(F) := Q_S^{(2)}(F)^2,$$

we obtain two SQM $(\mathfrak{F}, \Gamma_{bf}, Q_S^{(1)}(F), H_S^{(1)}(F))$ and $(\mathfrak{F}, \Gamma_{bf}, Q_S^{(2)}(F), H_S^{(2)}(F))$. We have

$$H_S(F) \subset H_S^{(a)}(F), \quad a = 1, 2.$$

A simple application of Theorem 1.7 yields:

Theorem 4.13

(i) *Suppose that, for some constant $\beta_0 > 0$, $e^{-\beta_0 H_S^{(1)}(F)}$ is trace class. Then, for all $\beta \geq \beta_0$, $e^{-\beta H_S^{(1)}(F)}$ is trace class and $Q_{S,+}(F)$ is Fredholm with*

$$\text{ind } Q_{S,+}(F) = \text{Tr}\left(\Gamma_{\text{bf}} e^{-\beta H_S^{(1)}(F)}\right),$$

independently of $\beta \geq \beta_0$.

(ii) *Suppose that, for some constant $\beta_0 > 0$, $e^{-\beta_0 H_S^{(2)}(F)}$ is trace class. Then, for all $\beta \geq \beta_0$, $e^{-\beta H_S^{(2)}(F)}$ is trace class and $Q_{S,-}(F)$ is Fredholm with*

$$\text{ind } Q_{S,-}(F) = -\text{Tr}\left(\Gamma_{\text{bf}} e^{-\beta H_S^{(2)}(F)}\right),$$

independently of $\beta \geq \beta_0$.[5]

4.11 Explicit Form of Supersymmetric Hamiltonian $H_S(F)$

In view of Theorem 4.13, we need to know an explicit form of $H_S^{(a)}(F)$ $(a = 1, 2)$. To find it, however, we need some conditions on F.

(F.2) For some $p > 4$, $\|F\|_{\mathscr{H}} \in L^p(Q_{\mathfrak{h}}, d\mu_{\mathfrak{h}})$, where, for $\alpha \geq 1$, $L^\alpha(Q_{\mathfrak{h}}, d\mu_{\mathfrak{h}}) := \{\Psi : Q_{\mathfrak{h}} \mapsto \mathbb{C} \cup \{\pm\infty\}$, measurable$| \int_{Q_{\mathfrak{h}}} |\Psi(q)|^\alpha d\mu_{\mathfrak{h}}(q) < \infty\}$.

(F.3) For a.e. q, $F(q) \in \text{Dom}(S^*)$ and $C S^* F(q) = S^* F(q)$, where C is the complex conjugation on $\mathfrak{h}_{\mathbb{C}}$. Moreover, $S^* F \in \text{Dom}(\nabla^*)$, where $(S^* F)(q) := S^* F(q)$ for a.e. q and ∇^* is the adjoint of the gradient operator ∇ (see (3.6) and (3.7)) and $\nabla^* S^* F \in L^r(Q_{\mathfrak{h}}, d\mu_{\mathfrak{h}})$ for some $r > 2$.

(F.4) For all $u \in \mathscr{H}$, the function $F_u(\cdot) := \langle F(\cdot), u \rangle_{\mathscr{H}}$ on $Q_{\mathfrak{h}}$ is in $L^{r'}(Q_{\mathfrak{h}}, d\mu_{\mathfrak{h}})$ for some $r' > 2$ and

$$F_u, \ F_u^* \in \text{Dom}(\nabla), \quad (\nabla F_u)(q), \ (\nabla F_u^*)(q) \in \text{Dom}(S) \quad (\text{a.e.} q).$$

Moreover, for a.e. q, the linear operator $T_F(q)$ and the anti-linear operator $R_F(q)$ on $\mathscr{H}$ defined by

$$T_F(q)u := \frac{1}{\sqrt{2}} S(\nabla F_u)(q), \quad R_F(q)u := \frac{1}{\sqrt{2}} S(\nabla F_u^*)(q), \quad u \in \mathscr{H}$$

are bounded and Hilbert–Schmidt respectively, and $\|T_F(\cdot)\|$, $\|R_F(\cdot)\|_2 \in L^s(Q_{\mathfrak{h}}, d\mu_{\mathfrak{h}})$ for some $s > 2$.

---

[5] Here we use the following fact also: for each Fredholm operator A, ind $A^* = -\text{ind } A$.

(F.5) There exists a sequence $\{F_N\}_{N=1}^{\infty}$ with $F_N \in \mathscr{P}(\mathrm{Dom}(S))\hat{\otimes}\mathrm{Dom}(S^*)$ such that, for some constants $p > 4, r > 2$ and $s > 2$,

$$\lim_{N\to\infty} \int_{Q_{\mathfrak{h}}} \|F_N - F\|_{\mathscr{K}}^p d\mu_{\mathfrak{h}}(q) = 0,$$

$$\lim_{N\to\infty} \int_{Q_{\mathfrak{h}}} |\nabla^* S^* F_N(q) - \nabla^* S^* F(q)|^r d\mu_{\mathfrak{h}}(q) = 0,$$

$$\lim_{N\to\infty} \int_{Q_{\mathfrak{h}}} \|T_{F_N}(q) - T_F(q)\|^s d\mu_{\mathfrak{h}}(q) = 0,$$

$$\lim_{N\to\infty} \int_{Q_{\mathfrak{h}}} \|R_{F_N}(q) - R_F(q)\|_2^s d\mu_{\mathfrak{h}}(q) = 0.$$

Remark 4.10 Condition (F.2) implies (F.1).

In Sect. 2.12, we have introduced fermion quadratic operators. These operators have extensions to the boson–fermion Fock space $\mathfrak{F}$ as shown below. For a $\mathfrak{B}(\mathscr{K})$-valued measurable function $T(\cdot)$ on $Q_{\mathfrak{h}}$, we define the operator $\langle B^*|T(\cdot)|B\rangle$ on $\mathfrak{F}$ as follows:

$$\mathrm{Dom}(\langle B^*|T(\cdot)|B\rangle) := \{\Psi \in \mathfrak{F} | \int_{Q_{\mathfrak{h}}} \|\langle B^*|T(q)|B\rangle\Psi(q)\|_{\mathscr{F}_{\mathrm{f}}(\mathscr{K})}^2 d\mu_{\mathfrak{h}}(q) < \infty\},$$

$$(\langle B^*|T(\cdot)|B\rangle\Psi)(q) := \langle B^*|T(q)|B\rangle\Psi(q), \quad \Psi \in \mathrm{Dom})\langle B^*|T(\cdot)|B\rangle), \ \text{a.e. } q.$$

If $T(q)$ is Hilbert–Schmidt or anti-linear Hilbert–Schmidt for a.e. q, then one can define also the operators $\langle B^*|T(\cdot)|B^*\rangle$ and $\langle B|T(\cdot)|B\rangle$ on $\mathfrak{F}$ as follows:

$$\mathrm{Dom}(\langle B^*|T(\cdot)|B^*\rangle) := \{\Psi \in \mathfrak{F} | \int_{Q_{\mathfrak{h}}} \|\langle B^*|T(q)|B^*\rangle\Psi(q)\|_{\mathscr{F}_{\mathrm{f}}(\mathscr{K})}^2 d\mu_{\mathfrak{h}}(q) < \infty\},$$

$$(\langle B^*|T(\cdot)|B^*\rangle\Psi)(q) := \langle B^*|T(q)|B^*\rangle\Psi(q), \quad \Psi \in \mathrm{Dom})\langle B^*|T(\cdot)|B^*\rangle), \ \text{a.e. } q,$$

$$\mathrm{Dom}(\langle B|T(\cdot)|B\rangle) := \{\Psi \in \mathfrak{F} | \int_{Q_{\mathfrak{h}}} \|\langle B|T(q)|B\rangle\Psi(q)\|_{\mathscr{F}_{\mathrm{f}}(\mathscr{K})}^2 d\mu_{\mathfrak{h}}(q) < \infty\},$$

$$(\langle B|T(\cdot)|B\rangle\Psi)(q) := \langle B|T(q)|B\rangle\Psi(q), \quad \Psi \in \mathrm{Dom})\langle B|T(\cdot)|B\rangle), \ \text{a.e. } q.$$

Proposition 4.4 *Assume (F.2)–(F.5). Then $\hat{\mathscr{D}}_S \subset \mathrm{Dom}(Q_S(F)^2) = \mathrm{Dom}(H_S(F))$ and*

$$H_S(F) = \Delta_S + \frac{1}{\sqrt{2}} \nabla^* S^* F + \|F(\cdot)\|_{\mathcal{H}}^2 + \langle B^* | T_F(\cdot) | B \rangle$$
$$+ \langle B^* | T_F(\cdot)^* | B \rangle + \langle B^* | R_F(\cdot) | B^* \rangle + \langle B | R_F(\cdot) | B \rangle . \qquad (4.49)$$

on $\hat{\mathcal{D}}_S$.

Proof We give only an outline of the proof. We have

$$Q_S(F)^2 \supset Q_S^2 + \{d_S, B(F)\} + \{d_S, B(F)^*\} + \{d_S^*, B(F)\} + \{d_S^*, B(F)^*\}$$
$$+ B(F)^2 + (B(F)^*)^2 + B(F)B(F)^* + B(F)^* B(F).$$

It follows from (2.19) that $B(F)^2 = 0$ and $(B(F)^*)^2 = 0$ on $\hat{\mathcal{D}}_S$ and

$$B(F)B(F)^* + B(F)^* B(F) = \|F(\cdot)\|_{\mathcal{H}}^2 \quad \text{on } \hat{\mathcal{D}}_S,$$

where we have used condition (F.2). We already know that $\text{Dom}(Q_S^2) \supset \hat{\mathcal{D}}_S$ and $Q_S^2 = \Delta_S$. Hence

$$Q_S(F)^2 \supset \Delta_S + \|F(\cdot)\|_{\mathcal{H}}^2 + \{d_S, B(F)\} + \{d_S, B(F)^*\} + \{d_S^*, B(F)\} + \{d_S^*, B(F)^*\}.$$

To proceed further, we first consider the case where $F \in \mathcal{P}(\text{Dom}(S)) \hat{\otimes} \text{Dom}(S^*)$. In this case, there exist $K \in \mathbb{N}$, $G_k \in \mathcal{P}(\text{Dom}(S))$ and $v_k \in \text{Dom}(S^*)$ ($k = 1, \ldots, K$) such that $F = \sum_{k=1}^{K} G_k \otimes v_k$. Let d be the dimension of the subspace span $\{v_1, \ldots, v_K\}$ and $\{e_n\}_{n=1}^{N}$ be a CONS of span $\{v_1, \ldots, v_K\}$. Then one can rewrite F in terms of $\{e_n\}_{n=1}^{N}$ to obtain $F = \sum_{n=1}^{N} H_n \otimes e_n$ with $H_n \in \mathcal{P}(\text{Dom}(S))$. It is not so difficult to show that $\hat{\mathcal{D}}_S$ is included in $\text{Dom}(\{d_S, B(F)\}) \cap \text{Dom}(\{d_S, B(F)^*\}) \cap \text{Dom}(\{d_S^*, B(F)\}) \cap \text{Dom}(\{d_S^*, B(F)^*\})$ and the following equations hold on $\hat{\mathcal{D}}_S$:

$$\{d_S, B(F)\} = \langle B^* | T_F(\cdot) | B \rangle + \frac{1}{\sqrt{2}} \sum_{n=1}^{N} \langle F, e_n \rangle D_{S^* e_n},$$

$$\{d_S, B(F)^*\} = \langle B^* | R_F(\cdot) | B^* \rangle$$

$$\{d_S^*, B(F)\} = \langle B | R_F(\cdot) | B \rangle,$$

$$\{d_S^*, B(F)^*\} = \left\langle B^* | T_F(\cdot)^* | B \rangle + \frac{1}{\sqrt{2}} \sum_{n=1}^{N} \right\rangle D_{S^* e_n}^* \cdot \langle e_n, F \rangle,$$

where we have formula (4.45) is used. By (3.4), we have

$$\frac{1}{\sqrt{2}} \sum_{n=1}^{N} D^*_{S^* e_n} \cdot \langle e_n, F \rangle = -\frac{1}{\sqrt{2}} \sum_{n=1}^{N} \langle e_n, F \rangle D_{CS^* e_n} + \frac{1}{\sqrt{2}} \sum_{n=1}^{N} (D^*_{S^* e_n} \langle e_n, F \rangle).$$

Condition (F.3) implies that $\sum_{n=1}^{N} \langle F, e_n \rangle D_{S^* e_n} = \sum_{n=1}^{N} \langle e_n, F \rangle D_{CS^* e_n}$ on $\hat{\mathcal{D}}_S$. One can show that $\sum_{n=1}^{N} \langle e_n, F \rangle D_{CS^* e_n} = \nabla^* S^* F$ on $\hat{\mathcal{D}}_S$. Thus we see that $\hat{\mathcal{D}}_S \subset$ Dom$(Q_S(F)^2)$ and (4.49) holds. Then, by a limiting argument using condition (F.5), one can extended the result to a general F obeying the assumption in Proposition 4.4. For more details, see the proof of [12, Theorem 4.4] (note that notations and methods there are slightly different from those of the present book). Cf. also [11]. $\square$

We formulate an additional condition under which $\langle B^* | R_F | B^* \rangle$ and $\langle B | R_F | B \rangle$ vanish on $\hat{\mathcal{D}}_S$:

(F.6) For all $u, v \in$ Dom(S^*) and a.e. $q \in Q_{\mathfrak{h}}$,

$$\langle S^* u, (\nabla F_v^*)(q) \rangle_{\mathfrak{h}_{\mathbb{C}}} = \langle S^* v, (\nabla F_u^*)(q) \rangle_{\mathfrak{h}_{\mathbb{C}}}.$$

Example 4.1 Let $W \in$ Dom(∇) such that, for a.e. q, $(\nabla W)(q) \in$ Dom(S). Then the $\mathcal{K}$-valued function $F := S \nabla W / \sqrt{2}$ on $Q_{\mathfrak{h}}$—see (4.43)—satisfies (F.6).

Lemma 4.16 *For a.e. q,*

$$\langle B^* | R_F(q) | B^* \rangle = 0 \quad on \ \mathcal{F}_{f, fin}(\mathcal{K})$$

if and only if (F.6) holds. In that case, we have

$$\langle B | R_F(q) | B \rangle = 0 \quad on \ \mathcal{F}_{f, fin}(\mathcal{K})$$

Proof Let $\Psi := B(u_1)^* \cdots B(u_p)^* \Omega_{\mathcal{K}}$ $(p \geq 0, u_1, \ldots, u_p \in \mathcal{K})$ and $\{e_n\}_{n=1}^{\infty}$ be a CONS of $\mathcal{K}$. Then

$$(\langle B^* | R_F(q) | B^* \rangle \Psi)^{(p+2)} = \sqrt{(p+2)!} A_{p+2}(T \otimes u_1 \otimes \cdots \otimes u_p),$$

where $T := \sum_{n=1}^{\infty} R_F(q) e_n \otimes e_n$. Hence $\langle B^* | R_F(q) | B^* \rangle \Psi = 0$ for all $u_1, \ldots, u_p \in \mathcal{K}$ if and only if $A_2(T) = 0$. On the other hand, $A_2(T) = 0$ is equivalent to (F.6). The second half of the lemma follows from that $\langle B | R_F | B \rangle = \langle B^* | R_F | B^* \rangle^*$ on $\hat{\mathcal{D}}_S$. $\square$

Lemma 4.16 implies that, under conditions (F.2)–(F.6), $H_S(F)$ takes the following simpler form on $\hat{\mathcal{D}}_S$:

$$H_S(F) = \Delta_S + \frac{1}{\sqrt{2}} \nabla^* S^* F + \| F(\cdot) \|_{\mathcal{K}}^2 + \langle B^* | T_F(\cdot) | B \rangle + \langle B^* | T_F(\cdot)^* | B \rangle. \quad (4.50)$$

4.12 Path Integral Representation of the Index of $Q_{S,+}(F)$

Under additional conditions, one can derive a path integral representation for the index ind $Q_{S,+}(F)$. In this section, we briefly describe this aspect.

We continue to assume (F.2)–(F.5).

4.12.1 Path Integral Representations of Pure Imaginary Time Correlation Functions of Bose Fields

An additional assumption is the following:

(S.1) The non-negative self-adjoint operators

$$h := S^*S$$

and SS^* are strictly positive and, for some constant $\gamma > 1$, $h^{-(\gamma-1)}$ is trace class on $\mathfrak{h}$.

Under Assumption (S.1), $h^{-(\gamma-1)}$ is a strictly positive and self-adjoint compact operator. Hence there exists a CONS $\{e_n\}_{n=1}^{\infty}$ of $\mathfrak{h}_{\mathbb{C}}$ and positive numbers ε_n such that $he_n = \varepsilon_n e_n, \varepsilon_n \leq \varepsilon_{n+1}$ $(n \geq 1)$ and $\operatorname{Tr} h^{-(\gamma-1)} = \sum_{n=1}^{\infty} \varepsilon_n^{-(\gamma-1)} < \infty$. It follows that $\sigma(h) = \sigma_p(h) = \{\varepsilon_n\}_{n=1}^{\infty}$ and $\sum_{n=1}^{\infty} \varepsilon_n^{-(\gamma-1)} < \infty$. In particular, we have

$$\operatorname{Tr} h^{-\gamma} = \sum_{n=1}^{\infty} \frac{1}{\varepsilon_n^{\gamma}} < \infty. \tag{4.51}$$

Hence $h^{-\gamma}$ is trace class on $\mathfrak{h}$.

The domain $\operatorname{Dom}(h^{\gamma/2}) \cap \mathfrak{h}$ is a real Hilbert space with the inner product $\langle f, g \rangle_{\gamma} := \langle h^{\gamma/2}f, h^{\gamma/2}g \rangle_{\mathfrak{h}}$ $(f, g \in \operatorname{Dom}(h^{\gamma/2}) \cap \mathfrak{h})$ and the norm $\|f\|_{\gamma} := \sqrt{\langle f, f \rangle_{\gamma}}$. We denote this Hilbert space by $\mathfrak{h}_{\gamma}$. On the other hand, $\mathfrak{h}$ is a real inner product space with the inner product $\langle f, g \rangle_{-\gamma} := \langle h^{-\gamma/2}f, h^{-\gamma/2}g \rangle$ $(f, g \in \mathfrak{h})$. We denote the completion of this inner product space by $\mathfrak{h}_{-\gamma}$. It follows that $\mathfrak{h}_{-\gamma}$ is the dual space of $\mathfrak{h}_{\gamma}$ with the natural bilinear form $\langle \varphi, f \rangle$ $(\varphi \in \mathfrak{h}_{-\gamma}, f \in \mathfrak{h}_{\gamma})$ such that $\langle \varphi, f \rangle = \langle \varphi, f \rangle_{\mathfrak{h}}$ if $\varphi \in \mathfrak{h}$ and, for all $\varphi \in \mathfrak{h}_{-\gamma}$ and $f \in \mathfrak{h}_{\gamma}, |\langle \varphi, f \rangle| \leq \|\varphi\|_{-\gamma} \|f\|_{\gamma}$, where $\|\cdot\|_{-\gamma}$ denotes the norm of $\mathfrak{h}_{-\gamma}$. By (S.1), the embedding of $\mathfrak{h}$ into $\mathfrak{h}_{-\gamma}$ is Hilbert–Schmidt. Hence, by the Minlos theorem (e.g., [20, Theorem D.18], [32], [38, Theorem 1.72]), we can take the measure space $Q_{\mathfrak{h}}$ to be $\mathfrak{h}_{-\gamma}$. Letting

$$E := \mathfrak{h}_{-\gamma}$$

we denote the measure $\mu_{\mathfrak{h}}$ on E by μ_E. In this case, we have $\varphi_{\mathfrak{h}}(f)(\varphi) = \langle \varphi, f \rangle$ $(\varphi \in E, f \in \mathfrak{h})$.[6]

---

[6] For $f \in \mathfrak{h}$ and $\varphi \in E$, $\langle \varphi, f \rangle := \lim_{n \to \infty} \langle \varphi, f_n \rangle$ in $L^2(E, d\mu_E)$, where $\{f_n\}_n$ is a sequence in $\mathfrak{h}_{\gamma}$ such that $\lim_{n \to \infty} f_n = f$ in $\mathfrak{h}$. The limit $\lim_{n \to \infty} \langle \varphi, f_n \rangle$ is independent of the choice of $\{f_n\}_n$.

Lemma 4.17 *For all* $\beta > 0$, $e^{-\beta h}$ *is trace class.*

Proof It is easy to see that $C := \sup_{x>0} x^{\gamma} e^{-\beta x} < \infty$. Hence $\sum_{n=1}^{\infty} \langle e_n, e^{-\beta h} e_n \rangle = \sum_{n=1}^{\infty} e^{-\beta \varepsilon_n} \leq C \sum_{n=1}^{\infty} \varepsilon_n^{-\gamma} < \infty$ (by (4.51)). □

Since $\|e^{-\beta h}\| < 1$ for all $\beta > 0$, it follows that $1 - e^{-\beta h}$ is strictly positive and bijective. In particular, the inverse $(1 - e^{-\beta h})^{-1}$ exists.

The boson second quantization operator $d\Gamma_b(h)$ physically denotes the Hamiltonian of a free Bose field with one-boson Hamiltonian h.

Lemma 4.18 *For all* $\beta > 0$, $e^{-\beta d\Gamma_b(h)}$ *is trace class and*

$$\mathrm{Tr}\, e^{-\beta d\Gamma_b(h)} = \frac{1}{\prod_{n=1}^{\infty}(1 - e^{-\beta \varepsilon_n})}.$$

Proof See [22, Corollary 5.5]. □

In the context of quantum statistical mechanics, the function

$$Z_\beta := \mathrm{Tr}\, e^{-\beta d\Gamma_b(h)}$$

of $\beta > 0$ (the inverse temperature) is called the **partition function** of the free Bose field system with the Hamiltonian $d\Gamma_b(h)$.

Let $\mathfrak{A}$ be a $*$-algebra consisting of linear operators on $\mathscr{F}_b(\mathfrak{h}_\mathbb{C})$ such that I (the identity)$\in \mathfrak{A}$ and, for all $A \in \mathfrak{A}$ and $t > 0$, $Ae^{-td\Gamma_b(h)}$ is trace class on $\mathscr{F}_b(\mathfrak{h}_\mathbb{C})$ or a densely defined bounded linear operator whose extension is trace class on $\mathscr{F}_b(\mathfrak{h}_\mathbb{C})$. Then one can define a linear functional $\omega_\beta : \mathfrak{A} \to \mathbb{C}$ by

$$\omega_\beta(A) := \frac{\mathrm{Tr}\,(Ae^{-\beta d\Gamma_b(h)})}{Z_\beta}, \quad A \in \mathfrak{A}.$$

It follows that $\omega_\beta(I) = 1$ and $\omega_\beta(A^*A) \geq 0$ for all $A \in \mathfrak{A}$. Hence ω_β is a state on $\mathfrak{A}$. It is called the **Gibbs state** associated with $d\Gamma_b(h)$.

The time development of the time-zero Bose field $\phi_C(f)$ ($f \in \mathfrak{h}$) under the Hamiltonian $d\Gamma_b(h)$ is defined by $e^{itd\Gamma_b(h)}\phi_C(f)e^{-itd\Gamma_b(h)}$, where $t \in \mathbb{R}$ is the time parameter. Replacing t by it, we see that it is natural to define the "pure imaginary time development" $\phi_C(it, f)$ of $\phi_C(f)$ by

$$\phi_C(it, f) := e^{-td\Gamma_b(h)}\phi_C(f)e^{td\Gamma_b(h)}.$$

For each $n \in \mathbb{N}$, the n-point correlation functions of pure imaginary time free Bose fields with respect to the Gibbs state ω_β are defined by

$$G_n(t_1, f_1; \cdots ; t_n, f_n) := \omega_\beta(\phi_C(it_1, f_1) \cdots \phi_C(it_n, f_n)),$$
$$0 \leq t_1 \leq t_2, \cdots \leq t_n < \beta, f_1, \ldots, f_n \in \mathfrak{h}.$$

The following lemma gives a trace formula for the two-point correlation functions:

Lemma 4.19 *Let $0 \leq s \leq t < \beta$ and $f, g \in \mathfrak{h}$. Then*

$$G_2(s, g; t, f) = \frac{1}{2}\langle g, (e^{-(t-s)h} + e^{-(\beta-(t-s))h})(1 - e^{-\beta h})^{-1} f \rangle_{\mathfrak{h}} . \tag{4.52}$$

Proof In the same way as in the proof of (2.14), one can show that

$$\phi_C(it, f) = \frac{1}{\sqrt{2}}(A(e^{-th} f)^* + A(e^{th} f)) \quad \text{on } \mathscr{F}_{b,\text{fin}}(\text{Dom}(e^{th})).$$

Hence

$$G_2(s, g; t, f) = \frac{1}{2}\{\omega_\beta(A(e^{-sh} g)^* A(e^{-th} f)^*) + \omega_\beta(A(e^{-sh} g)^* A(e^{th} f))$$
$$+ \omega_\beta(A(e^{sh} g)A(e^{-th} f)^*) + \omega_\beta(A(e^{sh} g)A(e^{th} f))\}.$$

It is not so difficult to show (see [29, §5.2], [19, Theorem 8.16]) that, for all $f_1, f_2 \in \mathfrak{h}_C$,

$$\omega_\beta(A(f_1)^* A(f_2)) = \langle f_2, e^{-\beta h}(1 - e^{-\beta h})^{-1} f_1 \rangle,$$
$$\omega_\beta(A(f_1)A(f_2)^*) = \langle f_1, (1 - e^{-\beta h})^{-1} f_2 \rangle,$$
$$\omega_\beta(A(f_1)^* A(f_2)^*) = 0, \quad \omega_\beta(A(f_1)A(f_2)) = 0.$$

Using these formulas, one obtains (4.52). $\square$

Remark 4.11 It is shown that, for all $n \geq 1$, $G_{2n-1} = 0$ and G_{2n} is written as a combinatorial sum of two-point functions G_2.

As in the case of Euclidean quantum field theory [61], one can represent the correlation function G_n in terms of functional integrations. A key fact for this is Lemma 4.20 below.

For each $\beta > 0$, we set

$$E_\beta := C([0, \beta]; E),$$

the space of continuous mappings from $[0, \beta]$ to E, which is a path space with paths in E. For each $\Phi \in E_\beta$, we set

$$\Phi_t := \Phi(t) \in E.$$

We denote by $\mathscr{F}$ the Borel field on E_β generated by $\{\Phi_t(f) | f \in \mathfrak{h}, t \in [0, \beta]\}$, where $\Phi_t(f) := \langle \Phi_t, f \rangle$.

Lemma 4.20 *There exist a probability measure μ_β on $(E_\beta, \mathscr{F})$ such that $\{\Phi_t(f)|f \in \mathfrak{h}, t \in [0, \beta]\}$ is a family of jointly Gaussian random variables on $(E_\beta, \mathscr{F}, \mu_\beta)$ with covariance*

$$\int_{E_\beta} d\mu_\beta \Phi_t(f)\Phi_s(g) = \frac{1}{2} \langle f, (e^{-|t-s|h} + e^{-(\beta-|t-s|)h})(1 - e^{-\beta h})^{-1}g \rangle_{\mathfrak{h}}, \quad (4.53)$$

for all $f, g \in \mathfrak{h}$ and $t, s \in [0, \beta]$.

Proof We need only apply the theory in Appendix B to the case $A = h$ (cf. [36, Proposition 5.1]). □

Remark 4.12 It follows from (4.53) that, for all $f \in \mathfrak{h}$, $\int_{E_\beta} |\Phi_0(f) - \Phi_\beta(f)|^2 d\mu_\beta = 0$. Hence $\Phi_0(f) = \Phi_\beta(f)$ for a.e. Φ. Since $\mathfrak{h}$ is separable, it follows that $\Phi_0 = \Phi_\beta$ for a.e. $\Phi \in E_\beta$. This means that, for a.e. Φ, $\Phi \in E_\beta$ is a loop in E. Hence one can regard the measure μ_β as the probability measure on the **loop space** $L([0, \beta]; E) := \{\Phi \in C([0, \beta]; E)|\Phi_0 = \Phi_\beta\}$.

We can now state formulas which represent correlation functions as functional integrals with respect to μ_β:

Theorem 4.14 *Let $f_1, \ldots, f_n \in \mathfrak{h}$, $0 < t_1 < t_2 < \cdots < t_n < \beta$. Then*

$$G_n(t_1, f_1; \cdots ; t_n, f_n) = \int_{E_\beta} \Phi_{t_1}(f_1) \cdots \Phi_{t_n}(f_n) d\mu_\beta(\Phi).$$

Proof See [15, Theorem 2.2].[7] □

Theorem 4.14 can be extended as follows. Let V be a real-valued measurable function on E which is finite for a.e. φ and bounded from below (i.e., there exists a real constant V_0 such that $V(\varphi) \geq V_0$ for a.e. φ) satisfying that $\text{Dom}(d\Gamma_{\mathfrak{b}}(h)^{1/2}) \cap \text{Dom}(\hat{V}^{1/2})$ is dense in $L^2(E, d\mu_E)$, where $\hat{V} := V - V_0 \geq 0$. Then, by the second representation theorem for densely defined closed symmetric forms [45, Chap. VI, Theorem 2.23], there exists a unique self-adjoint operator H_V on $L^2(E, d\mu_E)$, bounded from below, such that, for all $\Psi, \Theta \in \text{Dom}(d\Gamma_{\mathfrak{b}}(h)^{1/2}) \cap \text{Dom}(\hat{V}^{1/2})$,

$$\langle \hat{H}_V^{1/2}\Psi, \hat{H}_V^{1/2}\Theta \rangle = \langle d\Gamma_{\mathfrak{b}}(h)^{1/2}\Psi, d\Gamma_{\mathfrak{b}}(h)^{1/2}\Theta \rangle + \langle \hat{V}^{1/2}\Psi, \hat{V}^{1/2}\Theta \rangle,$$

where $\hat{H}_V := H_V - V_0 \geq 0$, and $\text{Dom}(\hat{H}_V^{1/2}) = \text{Dom}(d\Gamma_{\mathfrak{b}}(h)^{1/2}) \cap \text{Dom}(\hat{V}^{1/2})$. One can prove the following:

---

[7] Note that the convention on the Gaussian random process in [15] is different from that in the present book.

Theorem 4.15 *Let V and H_V be as above. Then:*

(i) *For all $\beta > 0$, $e^{-\beta H_V}$ is trace class.*
(ii) *Let $n \in \mathbb{N}$ and $F_1, \ldots, F_n$ be measurable functions on E such that, for all $t > 0$, $e^{-tH_V}|F_j|e^{-tH_V}$ ($j = 1, \ldots, n$) is trace class. Then, for all $t_1, \ldots, t_n \in [0, \beta]$ satisfying $0 < t_1 < t_2 < \cdots < t_n < \beta$,*

$$\frac{\text{Tr}\,(e^{-t_1 H_V} F_1 e^{-(t_2-t_1)H_V} \cdots F_{n-1} e^{-(t_n-t_{n-1})H_V} F_n e^{-(\beta-t_n)H_V})}{Z_\beta}$$

$$= \int_{E_\beta} F_1(\Phi_{t_1}) \cdots F_n(\Phi_{t_n}) e^{-\int_0^\beta V(\Phi_t)dt}\, d\mu_\beta.$$

Proof See [15, Theorem 3.1]. □

4.12.2 Index Formula

For a.e. $\varphi \in E$, the operator

$$L_F(\varphi) := T_F(\varphi) + T_F(\varphi)^*$$

is a self-adjoint Hilbert–Schmidt operator on $\mathscr{K}$. We denote by ∂_t the differential operator $\partial/\partial t$ with *periodic boundary condition* on $L^2([0, \beta]; \mathscr{K}) \cong L^2([0, \beta]) \otimes \mathscr{K}$. By using the unitary equivalence $\partial_t + SS^* \cong \partial_t \otimes I + I \otimes SS^*$, the theory of tensor product operators and the strict positivity of SS^*, one can show that $\partial_t + SS^*$ is bijective with $(\partial_t + SS^*)^{-1}$ being bounded. Hence, for a.e. $\Phi \in E_\beta$, we can define the bounded linear operator

$$K_F(\Phi) := L_F(\Phi_t)(\partial_t + SS^*)^{-1}$$

on $L^2([0, \beta]; \mathscr{K})$.

Lemma 4.21 *Suppose that*

$$\int_0^\beta \|L_F(\Phi_t)\|_2^2 dt < \infty, \quad a.e.\ \Phi. \tag{4.54}$$

Then, for a.e. Φ, $K_F(\Phi)$ is Hilbert–Schmidt on $L^2([0, \beta]; \mathscr{K})$ and there exists a bounded linear operator $K_F(\Phi; t, s)$ on $\mathscr{K}$ ($t, s \in [0, \beta], t \neq s$) such that the mapping $s \mapsto K_F(\Phi; t, s)$ ($s \neq t$) is strongly continuous and

$$(K_F(\Phi)f)(t) = \int_0^\beta K_F(\Phi; t, s) f(s) ds, \quad f \in L^2([0, \beta]; \mathscr{K}).$$

Moreover, $K_F(\Phi; t) := K_F(\Phi; t, t + 0)$ is trace class on $\mathscr{K}$ and

$$\widetilde{\mathrm{Tr}}\, K_F(\Phi) := \int_0^\beta \mathrm{Tr}\, K_F(\Phi; t) dt$$

is a real-valued measurable function on E_β.

Proof See [9, Lemma 6.6, Lemma 6.7].   □

Let

$$W_F := \frac{1}{\sqrt{2}} \nabla^* S^* F + \|F\|_{\mathscr{K}}^2.$$

Theorem 4.16 *Assume (F.2)–(F.6), (4.54) and the following:*

(i) *The operator $H_S(F)$ given by (4.50) is essentially self-adjoint on $\hat{\mathscr{D}}_S$ and, for all $\beta > 0$, $e^{-\beta \overline{H_S(F)}}$ is trace class.*
(ii) *The operator $d\Gamma_b(h) + W_F$ is essentially self-adjoint, bounded from below and, for all $\beta > 0$, $e^{-\beta(\overline{d\Gamma_b(h)+W_F})}$ is trace class.*
(iii)

$$\int_{E_\beta} d\mu_\beta \exp\left(-\int_0^\beta W_F(\Phi_s) ds + \frac{1}{2}\|K_F\|_2^2 + |\widetilde{\mathrm{Tr}}\, K_F|\right) < \infty.$$

Then $Q_{S,+}(F)$ is Fredholm and

$$\mathrm{ind}\, Q_{S,+}(F) = \int_{E_\beta} d\mu_\beta(\Phi) \det_2(1 + K_F(\Phi)) \exp\left(-\int_0^\beta W(\Phi_s) ds + \widetilde{\mathrm{Tr}}\, K_F(\Phi)\right)$$

$$(4.55)$$

independently of $\beta > 0$, where $\det_2(1 + \cdot)$ denotes the regularized determinant for a Hilbert–Schmidt operator (see Sect. 2.7).

Proof See [9, Theorem 6.8]. We remark that, by using suitable approximate and limiting arguments, one can remove some of the assumed conditions in [9, Theorem 6.8].   □

Remark 4.13 In the case where (F.6) does not hold, the operators $\langle B^* | R_F(\cdot) | B^* \rangle$ and $\langle B | R_F(\cdot) | B \rangle$ do not vanish. But, in this case too, one can derive a path integral representation for ind $Q_{S,+}(F)$ [9, Theorem 6.10] which is given by (4.55) with the factor $\det_2(1 + K_F(\Phi)) e^{\widetilde{\mathrm{Tr}}\, K_F(\Phi)}$ replaced by $\widetilde{\mathrm{Pf}}(A_S, B_F(\Phi)) \exp\left(-\frac{1}{2} \int_0^\beta \mathrm{Tr}\, K_F(\Phi_s)\, ds\right)$, where A_S and $B_F(\Phi)$ are bounded skew-symmetric operators on $\mathscr{K} \oplus \mathscr{K}$ (A_S does not depend on Φ) and $\widetilde{\mathrm{Pf}}(\cdot, \cdot)$ denotes the extended relative Pfaffian defined for a pair of bounded skew-symmetric operators [9, Appendix E].

We present, in the next chapter, examples of the infinite dimensional Dirac operator $Q_S(F)$ which appear in supersymmetric quantum field theory.

Chapter 5
Models in Supersymmetric Quantum Field Theory

Abstract We construct in a mathematically rigorous way two-dimensional versions of the Wess–Zumino models in supersymmetric quantum field theory as an application of the theory of infinite-dimensional Dirac operators on the abstract boson–fermion Fock space presented in Chap. 4. Other supersymmetric quantum field models also are briefly mentioned.

5.1 Preliminaries

We construct supersymmetric quantum field models on the two-dimensional (cylindrical) space–time

$$M_\ell := \mathbb{R} \times T_\ell = \{(t, x) | t \in \mathbb{R}, x \in T_\ell\}$$

with $T_\ell := \mathbb{R}/\ell\mathbb{Z}$ (one-torus), which is identified with the circle of length ℓ. For this purpose, we need some preparations.

5.1.1 Momentum Operator of a Quantum Particle in T_ℓ

In what follows, we use the physical unit system where the light speed c and the reduced Planck constant $\hbar$ are equal to 1. The momentum space of a quantum particle being in T_ℓ is given by

$$\hat{T}_\ell := \frac{2\pi\mathbb{Z}}{\ell}.$$

We denote by $\ell^2(\hat{T}_\ell)$ the Hilbert space of absolutely square summable functions on $\hat{T}_\ell$:

$$\ell^2(\hat{T}_\ell) := \left\{ u : \hat{T}_\ell \to \mathbb{C} \mid \sum_{p \in \hat{T}_\ell} |u(p)|^2 < \infty \right\}$$

with inner product $\langle u, v \rangle := \sum_{p \in \hat{T}_\ell} u(p)^* v(p)$, $u, v \in \ell^2(\hat{T}_\ell)$. We set

$$\mathcal{H}_\ell := \ell^2(\hat{T}_\ell).$$

For each $p \in \hat{T}_\ell$, we define a function e_p on T_ℓ by

$$e_p(x) := \frac{1}{\sqrt{\ell}} e^{ipx}, \quad x \in T_\ell.$$

It is well known that $\{e_p\}_{p \in \hat{T}_\ell}$ is a CONS of $L^2(T_\ell)$. One can show [23, Theorem 1.33] that there exists a unique unitary operator $\mathscr{F}_\ell : L^2(T_\ell) \to \mathcal{H}_\ell$ such that

$$(\mathscr{F}_\ell f)(p) = \langle e_p, f \rangle_{L^2(T_\ell)} = \frac{1}{\sqrt{\ell}} \int_{T_\ell} e^{-ipx} f(x) dx, \quad f \in L^2(T_\ell), \ p \in \hat{T}_\ell.$$

The unitary operator $\mathscr{F}_\ell$ is called the **one-dimensional discrete Fourier transform** associated with T_ℓ. We write

$$\hat{f} := \mathscr{F}_\ell f, \quad f \in L^2(T_\ell).$$

We define an operator $\hat{p}_\ell$ on $L^2(T_\ell)$ as follows:

$$\text{Dom}(\hat{p}_\ell) := \left\{ f \in L^2(T_\ell) \mid \sum_{p \in \hat{T}_\ell} |p \hat{f}(p)|^2 < \infty \right\},$$

$$\hat{p}_\ell f := \sum_{p \in \hat{T}_\ell} p \hat{f}(p) e_p, \quad f \in \text{Dom}(\hat{p}_\ell).$$

It is easy to see that, for each $p \in \hat{T}_\ell$, $e_p \in \text{Dom}(\hat{p}_\ell)$ and $\hat{p}_\ell e_p = p e_p$, i.e., p is an eigenvalue of $\hat{p}_\ell$ and e_p is an eigenvector of $\hat{p}_\ell$ belonging to p. Since $\{e_p\}_{p \in \hat{T}_\ell}$ is a CONS of $L^2(T_\ell)$, it follows that $\hat{p}_\ell$ is self-adjoint and $\sigma(\hat{p}_\ell) = \sigma_p(\hat{p}_\ell) = \hat{T}_\ell$. The operator $\hat{p}_\ell$ is called the **momentum operator with the periodic boundary condition** acting in $L^2(T_\ell)$. We have

$$\mathscr{F}_\ell \hat{p}_\ell \mathscr{F}_\ell^{-1} = p, \tag{5.1}$$

where the right-hand side denotes the multiplication operator by the funcion p on $\hat{T}_\ell$. Since $\hat{p}_\ell$ is self-adjoint, its square

$$-\Delta_\ell := \hat{p}_\ell^2$$

is a non-negative self-adjoint operator on $L^2(T_\ell)$. The operator Δ_ℓ is called the **Laplacian with the periodic boundary condition** acting in $L^2(T_\ell)$. The Hamiltonian of a relativistic free particle of mass $m > 0$ in T_ℓ is defined by

$$h_\ell := (\hat{p}_\ell^2 + m^2)^{1/2} = (-\Delta_\ell + m^2)^{1/2} \tag{5.2}$$

acting in $L^2(T_\ell)$. Let $\omega : \hat{T}_\ell \to \mathbb{R}$ be defined by

$$\omega(p) := \sqrt{p^2 + m^2}, \quad p \in \hat{T}_\ell. \tag{5.3}$$

Then, by (5.1) and the functional calculus, we have

$$\mathscr{F}_\ell h_\ell \mathscr{F}_\ell^{-1} = \omega,$$

where the right-hand side denotes the multiplication operator by the function ω. The value $\omega(p)$ denotes the energy of the relativistic free particle with momentum $p \in \hat{T}_\ell$.

5.1.2 The Free Quantum Klein–Gordon Field on M_ℓ

We now move to the boson Fock space $\mathscr{F}_b(\mathscr{H}_\ell)$ over $\mathscr{H}_\ell$ in the momentum representation. We denote the annihilation operator on $\mathscr{F}_b(\mathscr{H}_\ell)$ with test vector $u \in \mathscr{H}_\ell$ by $a(u)$. It is easy to see that, for each $f \in L^2(T_\ell)$ and $t \in \mathbb{R}$, $e^{it\omega}\hat{f}/\sqrt{\omega}$ is in $\mathscr{H}_\ell$. Therefore one can define

$$\phi(t, f) := \frac{1}{\sqrt{2}}\left\{a\left(\frac{e^{it\omega}\hat{f}}{\sqrt{\omega}}\right)^* + a\left(\frac{e^{it\omega}\widehat{f^*}}{\sqrt{\omega}}\right)\right\},$$

where f^* means the complex conjugate of f and $\widehat{f^*} := \mathscr{F}_\ell f^*$. In the same way as in [22, Theorem 5.31], one can show that, for all $f \in \mathrm{Dom}(\Delta_\ell)$ and $\Psi \in \mathscr{F}_{b,0}(\mathscr{H}_\ell)$, the correspondence $\mathbb{R} \ni t \mapsto \phi(t, f)\Psi$ is twice strongly differentiable and the following equation holds:

$$\frac{d^2}{dt^2}\phi(t, f)\Psi + \phi(t, (-\Delta_\ell + m^2)f)\Psi = 0,$$

where d/dt means strong differentiation with respect to t. Thus the operator-valued functional $(t, f) \in \mathbb{R} \times \mathrm{Dom}(\Delta_\ell) \mapsto \phi(t, f)$ satisfies the free Klein–Gordon equation on the subspace $\mathscr{F}_{b,0}(\mathscr{H}_\ell)$. Moreover, if $f = f^*$, $f \in L^2(T_\ell)$, then $\phi(t, f)$ is the Segal field operator with test vector $e^{it\omega}\hat{f}/\sqrt{\omega}$. Hence it is symmetric and essentially self-adjoint (see Sect. 2.9). In particular, $\phi(t, f)$ is a neutral Bose field. The operator-valued functional $\phi : (t, f) \mapsto \phi(t, f)$ is called the **free quantum Klein–Gordon field** on M_ℓ. Obviously the time-zero field of ϕ takes the form

$$\phi(f) := \phi(0, f) = \frac{1}{\sqrt{2}} \left\{ a \left(\frac{\hat{f}}{\sqrt{\omega}} \right)^* + a \left(\frac{\widehat{f^*}}{\sqrt{\omega}} \right) \right\}, \quad f \in L^2(T_\ell). \tag{5.4}$$

The canonical conjugate momentum operator of $\phi(t, f)$ is defined by

$$\pi(t, f) := \frac{i}{\sqrt{2}} \left\{ a \left(\sqrt{\omega} e^{it\omega} \hat{f} \right)^* - a \left(\sqrt{\omega} e^{it\omega} \widehat{f^*} \right) \right\}$$

for all $f \in L^2(T_\ell)$ such that $\sqrt{\omega} \hat{f} \in \mathscr{H}_\ell$. One can easily show that, for each $t \in \mathbb{R}$, $\{\phi(t, f), \pi(t, g) | f, g \in L^2(T_\ell), \sqrt{\omega} \hat{g} \in \mathscr{H}_\ell\}$ obeys the Heisenberg CCR on $\mathscr{F}_{b,0}(\mathscr{H}_\ell)$: for all $f_1, f_2, g_1, g_2 \in L^2(T_\ell)$ such that $\sqrt{\omega} \hat{g}_1, \sqrt{\omega} \hat{g}_2 \in \mathscr{H}_\ell$,

$$[\phi(t, f_1), \pi(t, g_1)] = i \int_{T_\ell} f_1(x) g_1(x) dx, \tag{5.5}$$

$$[\phi(t, f_1), \phi(t, f_2)] = 0, \quad [\pi(t, g_1), \pi(t, g_2)] = 0 \tag{5.6}$$

on $\mathscr{F}_{b,0}(\mathscr{H}_\ell)$.

To understand a mathematical feature of $\phi(t, f)$ and $\pi(t, f)$, we recall a notion. We denote by $\mathscr{D}(T_\ell)$ the Fréchet space of infinitely differentiable functions on T_ℓ with the family $\{\rho_n\}_{n=0}^\infty$ of semi-norms defined by

$$\rho_n(f) := \sup_{x \in T_\ell} \left| \frac{d^n f(x)}{dx^n} \right|, \quad n \geq 0.$$

The Fréchet space $\mathscr{D}(T_\ell)$ is called the space of test functions on T_ℓ. The space of continuous linear functionals on $\mathscr{D}(T_\ell)$), denoted by $\mathscr{D}'(T_\ell)$, is called the space of distributions on T_ℓ (cf. [58, Chap. IV, §IV.2]). A distribution F on T_ℓ is said to be real if $F(f)^* = F(f^*)$ for all $f \in \mathscr{D}(T_\ell)$. We denote the space of real distributions on T_ℓ by $\mathscr{D}'_{\text{real}}(T_\ell)$. Let $\mathscr{D}_{\text{real}}(T_\ell)$ be the space of real-valued test functions on T_ℓ:

$$\mathscr{D}_{\text{real}}(T_\ell) := \{f \in \mathscr{D}(T_\ell) | f = f^*\}.$$

Then, for all $f \in \mathscr{D}_{\text{real}}(T_\ell)$ and $F \in \mathscr{D}'_{\text{real}}(T_\ell)$, $F(f)$ is a real number.

Let $\mathscr{K}$ be a Hilbert space and $\mathscr{D}$ be a dense subspace of $\mathscr{K}$. Suppose that, for each $f \in \mathscr{D}(T_\ell)$, a linear operator $F(f)$ on $\mathscr{K}$ is given such that, for all $f \in \mathscr{D}(T_\ell)$, $\text{Dom}(F(f)) \subset \mathscr{D}$ and, for all $\Psi \in \mathscr{D}$, $F(f)\Psi$ is linear in f. The subspace $\mathscr{D}$ is called a common domain for $F(f)$, $f \in \mathscr{D}(T_\ell)$. If, for all $\Psi, \Phi \in \mathscr{D}$, the functional $\langle \Psi, F(f)\Phi \rangle$ of $f \in \mathscr{D}(T_\ell)$ is continuous in f, i.e., the correspondence $f \mapsto \langle \Psi, F(f)\Phi \rangle$ is an element of $\mathscr{D}'(T_\ell)$, then F is called an **operator-valued distribution** on T_ℓ. In this case, one often introduces a symbol $F(x)$ (which has no mathematical meaning a priori) to write formally $F(f) = \int_{T_\ell} F(x)f(x)dx$ and calls $F(x)$ the distribution kernel of F.

A basic fact is that, for each $t \in \mathbb{R}$, $\phi(t, \cdot)$ and $\pi(t, \cdot)$ are operator-valued distributions on T_ℓ with $\mathscr{F}_{b,0}(\mathscr{H})$ being a common domain. Hence one can write formally

$$\phi(t, f) = \int_{T_\ell} \phi(t, x)f(x)dx, \quad \pi(t, f) = \int_{T_\ell} \pi(t, x)f(x)dx$$

with operator-valued distribution kernels $\phi(t, x)$ and $\pi(t, x)$.

Let

$$H_b := d\Gamma_b(\omega), \tag{5.7}$$

the boson second quantization operator of the multiplication operator ω acting in $\mathscr{H}_\ell$. Then, by (2.16), we have

$$\phi(t, f) = e^{itH_b}\phi(f)e^{-itH_b}, \quad t \in \mathbb{R}, \quad f \in L^2(T_\ell).$$

Thus H_b may be interpreted as the Hamiltonian of the quantum system described by $\{\phi(t, f)|t \in \mathbb{R}, f \in L^2(T_\ell)\}$.

For each $p \in \hat{T}_\ell$, we define $\delta_p : \hat{T}_\ell \to \mathbb{R}$ by $\delta_p(q) := \delta_{pq}$, $q \in \hat{T}_\ell$ with δ_{pq} being the Kronecker delta. Then it is obvious that $\delta_p \in \mathscr{H}_\ell$. Hence one can define the operator

$$a(p) := a(\delta_p).$$

We call it the **boson annihilation operator with momentum** p. It follows that

$$[a(p), a(q)^*] = \delta_{pq}, \quad [a(p), a(q)] = 0, \quad [a(p)^*, a(q)^*] = 0$$

on $\mathscr{F}_{b,0}(\mathscr{H}_\ell)$. Since $\{\delta_p | p \in \hat{T}_\ell\}$ is a CONS of $\mathscr{H}_\ell$, it follows that, for all $u \in \mathscr{H}_\ell$, $\lim_{N \to \infty} \sum_{p=-N}^{N} u(p)\delta_p = u$ in $\mathscr{H}_\ell$. Hence, by Lemma 2.1, for all $\Psi \in \mathscr{F}_{b,0}(\mathscr{H}_\ell)$,

$$a\left(\frac{e^{it\omega}\widehat{f^*}}{\sqrt{\omega}}\right)\Psi = \sum_{p \in \hat{T}_\ell} \frac{e^{-it\omega(p)}}{\sqrt{\omega(p)}} \hat{f}(-p)a(p)\Psi,$$

$$a\left(\frac{e^{it\omega}\hat{f}}{\sqrt{\omega}}\right)^*\Psi = \sum_{p \in \hat{T}_\ell} \frac{e^{it\omega(p)}}{\sqrt{\omega(p)}} \hat{f}(p)a(p)^*\Psi,$$

Therefore

$$\phi(t, f) := \sum_{p \in \hat{T}_\ell} \frac{1}{\sqrt{2\omega(p)}} \left(a(p)^* e^{it\omega(p)} \hat{f}(p) + a(p) e^{-it\omega(p)} \hat{f}(-p) \right) \quad \text{on } \mathscr{F}_{b,0}(\mathscr{H}_\ell).$$

This shows that the formal expression

$$\phi(t, x) = \frac{1}{\sqrt{\ell}} \sum_{p \in \hat{T}_\ell} \frac{1}{\sqrt{2\omega(p)}} (a(p)^* e^{it\omega(p)-ipx} + a(p) e^{-it\omega(p)+ipx})$$

may serve as the symbol for the operator-valued distribution kernel of $\phi(t, f)$, $f \in \mathscr{D}(T_\ell)$.[1] Similarly, the symbol for the operator-valued distribution kernel of $\pi(t, f)$, $f \in \mathscr{D}(T_\ell)$ is given by

$$\pi(t, x) = \frac{i}{\sqrt{\ell}} \sum_{p \in \hat{T}_\ell} \frac{\sqrt{\omega(p)}}{\sqrt{2}} (a(p)^* e^{it\omega(p)-ipx} - a(p) e^{-it\omega(p)+ipx}).$$

The Heisenberg CCR (5.5) and (5.6) are symbolically written as follows:

$$[\phi(t, x), \pi(t, y)] = i\delta(x - y),$$
$$[\phi(t, x), \phi(t, y)] = 0, \quad [\pi(t, x), \pi(t, y)] = 0,$$

where $\delta(x - y)$ is the delta-distribution on $T_\ell \times T_\ell$.

It is shown [22, Proposition 9.12] that the Hamiltonian H_b is written as

$$H_b = \sum_{p \in \hat{T}_\ell} \omega(p) a(p)^* a(p) \quad \text{on Dom}(H_b).$$

5.1.3 The Majorana Field on M_ℓ

The classical free Dirac equation in the space–time M_ℓ takes the following form:

$$(i\gamma^0 \partial_t + i\gamma^1 \partial_x - m)\psi^{cl}(t, x) = 0, \quad (t, x) \in M_\ell \tag{5.8}$$

with

$$\psi^{cl}(t, x) := \begin{pmatrix} \psi_+^{cl}(t, x) \\ \psi_-^{cl}(t, x) \end{pmatrix} \in \mathbb{C}^2,$$

---

[1] The symbol $\phi(t, x)$ here does not have meaning *as an operator* on $\mathscr{F}_b(\mathscr{H}_\ell)$. But one can give mathematical meaning to $\phi(t, x)$ as a sesquilinear form on a suitable domain (cf. [22, §9.6]).

being a continuously differentiable spinor on M_ℓ, where γ^0 and γ^1 are the gamma matrices associated with the two-dimensional Minkowski space-time, i.e., γ^0 and γ^1 are 2×2 Hermitian and anti-Hermitian matrices respectively satisfying the anti-commutation relations

$$\{\gamma^\mu, \gamma^\nu\} = 2g^{\mu\nu}, \quad \mu, \nu = 0, 1, \quad g^{00} = 1, g^{11} = -1, g^{01} = g^{10} = 0,$$

$m > 0$ is a parameter denoting physically the mass of the classical Dirac field ψ^{cl}, and $\partial_t := \partial/\partial t, \partial_x := \partial/\partial x$. Let

$$\alpha^1 := \gamma^0 \gamma^1.$$

Then, using the one-dimensional Dirac operator

$$h_D := -i\alpha^1 \partial_x + \gamma^0 m,$$

one can rewrite (5.8) as

$$i\partial_t \psi^{\text{cl}}(t, x) = h_D \psi^{\text{cl}}(t, x). \tag{5.9}$$

A solution ψ^{cl} of (5.8) is called a classical Majorana field if it is real: $\psi^{\text{cl}}(t, x)^* = \psi^{\text{cl}}(t, x), (t, x) \in M_\ell$. We first construct a classical Majorana field. For this purpose, it is convenient to use the following representation of the gamma matrices, called the Majorana representation:

$$\gamma^0 = \begin{pmatrix} 0 & -i \\ i & 0 \end{pmatrix}, \quad \gamma^1 = \begin{pmatrix} 0 & i \\ i & 0 \end{pmatrix}.$$

In this representation, we have

$$\alpha^1 = \begin{pmatrix} 1 & 0 \\ 0 & -1 \end{pmatrix}.$$

Hence

$$h_D = \begin{pmatrix} -i\partial_x & -im \\ im & i\partial_x \end{pmatrix},$$

We write the discrete Fourier transform of $\psi^{\text{cl}}(t, \cdot)$ as $\hat{\psi}^{\text{cl}}(t, \cdot)$:

$$\hat{\psi}^{\text{cl}}(t, p) := \frac{1}{\sqrt{\ell}} \int_{T_\ell} \psi^{\text{cl}}(t, x) e^{-ipx} dx, \quad p \in \hat{T}_\ell.$$

Then (5.9) is equivalent to

$$i\frac{d}{dt}\hat{\psi}^{\text{cl}}(t, p) = \hat{h}_D(p)\hat{\psi}^{\text{cl}}(t, p), \quad p \in \hat{T}_\ell \tag{5.10}$$

where

$$\hat{h}_D(p) := \alpha^1 p + \gamma^0 m = \begin{pmatrix} p & -im \\ im & -p \end{pmatrix}.$$

Since $\hat{h}_D(p)$ is Hermitian, it follows from (5.9) that

$$\hat{\psi}^{\text{cl}}(t, p) = e^{-it\hat{h}_D(p)}\hat{\psi}(0, p).$$

Thus the initial value problem of (5.10) (hence of (5.9)) is solved.

To find an explicit representation of $\hat{\psi}^{\text{cl}}(t, p)$, we need to solve the eigenvalue problem of $\hat{h}_D(p)$. But this is easy. The eigenvalues of $\hat{h}_D(p)$ are $\pm\omega(p)$ and normalized eigenvectors are given by

$$u_+(p) := \frac{1}{\sqrt{2\omega(p)}}\begin{pmatrix} v(p) \\ iv(-p) \end{pmatrix}, \quad u_-(p) := \frac{1}{\sqrt{2\omega(p)}}\begin{pmatrix} v(-p) \\ -iv(p) \end{pmatrix},$$

where $\omega(p)$ is defined by (5.3) and

$$v(p) := \sqrt{\omega(p) + p}.$$

Using the equation

$$v(p)v(-p) = m,$$

one can easily check that

$$\hat{h}_D(p)u_\pm(p) = \pm\omega(p)u_\pm(p).$$

Note that, for each $p \in \hat{T}_\ell$, $\{u_\pm(p)\}$ is an orthonormal basis of $\mathbb{C}^2$. Therefore

$$\hat{\psi}^{\text{cl}}(t, p) = e^{-it\omega)p)}c_+(p)u_+(p) + e^{it\omega(p)}c_-(p)u_-(p),$$

with $c_\pm(p) := \langle u_\pm(p), \hat{\psi}^{\text{cl}}(0, p)\rangle_{\mathbb{C}^2}$. Thus

$$\psi^{\text{cl}}(t, x) = \frac{1}{\sqrt{\ell}}\sum_{p\in\hat{T}_\ell} e^{ipx}\left\{e^{-it\omega(p)}c_+(p)u_+(p) + e^{it\omega(p)}c_-(p)u_-(p)\right\}.$$

Using the easily proved properties

$$u_+(-p)^* = u_-(p), \quad u_+(p) = u_-(-p)^*,$$

we see that the reality condition is equivalent to $c_-(p) = c_+(-p)^*$. Hence, putting $b^{\text{cl}}(p) := c_+(p)$, we find that the classical Majorana field is of the form

$$\psi^{\text{cl}}(t, x) = \frac{1}{\sqrt{\ell}} \sum_{p \in \hat{T}_\ell} e^{ipx} \left\{ e^{-it\omega)p)} u_+(p) b^{\text{cl}}(p) + e^{it\omega(p)} u_-(p) b^{\text{cl}}(-p)^* \right\}.$$

In components, we have

$$\psi^{\text{cl}}_+(t, x) = \frac{1}{\sqrt{\ell}} \sum_{p \in \hat{T}_\ell} \frac{e^{ipx}}{\sqrt{2\omega(p)}} \left\{ e^{-it\omega(p)} v(p) b^{\text{cl}}(p) + e^{it\omega(p)} v(-p) b^{\text{cl}}(-p)^* \right\},$$

(5.11)

$$\psi^{\text{cl}}_-(t, x) = \frac{i}{\sqrt{\ell}} \sum_{p \in \hat{T}_\ell} \frac{e^{ipx}}{\sqrt{2\omega(p)}} \left\{ e^{-it\omega(p)} v(-p) b^{\text{cl}}(p) - e^{it\omega(p)} v(p) b^{\text{cl}}(-p)^* \right\}.$$

(5.12)

To construct a quantum field version of $\psi^{\text{cl}}(t, x)$, we move to the fermion Fock space $\mathscr{F}_{\text{f}}(\mathscr{H}_\ell)$ over $\mathscr{H}_\ell$. We denote the fermion annihilation operator on $\mathscr{F}_{\text{f}}(\mathscr{H}_\ell)$ by $b(u)$, $u \in \mathscr{H}_\ell$. By (5.11) and (5.12), it is natural to define a two-component quantum field

$$\psi(t, f) = \begin{pmatrix} \psi_+(t, f) \\ \psi_-(t, f) \end{pmatrix}, \quad (t, f) \in \mathbb{R} \times L^2(T_\ell)$$

by

$$\psi_+(t, f) := \frac{1}{\sqrt{2}} \left[b\left(\frac{v e^{it\omega} \hat{f}}{\sqrt{\omega}} \right)^* + b\left(\frac{v e^{it\omega} \widehat{f^*}}{\sqrt{\omega}} \right) \right],$$

$$\psi_-(t, f) := -\frac{i}{\sqrt{2}} \left[b\left(\frac{\tilde{v} e^{it\omega} \hat{f}}{\sqrt{\omega}} \right)^* - b\left(\frac{\tilde{v} e^{it\omega} \widehat{f^*}}{\sqrt{\omega}} \right) \right],$$

where

$$\tilde{v}(p) := v(-p), \quad p \in \hat{T}_\ell.$$

Note that $\psi_\pm(t, f)$ are bounded operators. It follows that $\psi_\pm(t, \cdot)$ are operator-valued distributions on T_ℓ. If $f = f^*$, then they are bounded self-adjoint operators. Moreover, $\psi(t, f)$ satisfies the distributional Dirac equation

$$i\gamma^0 \partial_t \psi(t, f) + i\gamma^1 \psi(t, -\partial_x f) - m\psi(t, f) = 0, \quad (t, f) \in \mathbb{R} \times C^1(T_\ell),$$

where ∂_t means strong differentiation in t. It is easy to see that $\{\psi_\pm(t, f)|f \in L^2(T_\ell)\}$ obeys the CAR:

$$\{\psi_\alpha(t, f), \psi_\beta(t, g)\} = \delta_{\alpha\beta} \int_{T_\ell} f(x)g(x)dx, \quad f, g \in L^2(T_\ell), \ t \in \mathbb{R}, \ \alpha, \beta = \pm.$$

(5.13)

Thus $\psi(t, f)$ is a Majorana field on M_ℓ.

Let

$$H_f := d\Gamma_f(\omega),$$
(5.14)

the fermion second quantization operator of the multiplication operator ω acting in $\mathcal{H}_\ell$. Then, by (2.23), we have

$$\psi_\alpha(t, f) = e^{itH_f}\psi_\alpha(f)e^{-itH_f}, \quad t \in \mathbb{R}, \ f \in \mathcal{H}_\ell, \ \alpha = \pm,$$

where $\psi_\alpha(f) := \psi_\alpha(0, f) \ (\alpha = \pm)$ are the components of the time-zero Majorana field $\psi(0, f)$. Thus H_f may be interpreted as the Hamiltonian of the quantum system described by $\{\psi_\pm(t, f)|t \in \mathbb{R}, \ f \in \mathcal{H}_\ell\}$.

For each $p \in \hat{T}_\ell$, one can define the operator

$$b(p) := b(\delta_p).$$

We call it the **fermion annihilation operator with momentum** p. It follows that

$$\{b(p), b(q)^*\} = \delta_{pq}, \quad \{b(p), b(q)\} = 0, \quad \{b(p)^*, b(q)^*\} = 0.$$

As in the bosonic case, we can show that, for all $u \in \mathcal{H}_\ell$

$$b(u) = \sum_{p \in \hat{T}_\ell} u(p)^*b(p), \quad b(u)^* = \sum_{p \in \hat{T}_\ell} u(p)b(p)^*.$$

Therefore

$$\psi_+(t, f) = \sum_{p \in \hat{T}_\ell} \frac{v(p)}{\sqrt{2\omega(p)}} \left(e^{it\omega(p)}\hat{f}(p)b(p)^* + e^{-it\omega(p)}\hat{f}(-p)b(p)\right),$$

$$\psi_-(t, f) = -i \sum_{p \in \hat{T}_\ell} \frac{v(-p)}{\sqrt{2\omega(p)}} \left(e^{it\omega(p)}\hat{f}(p)b(p)^* - e^{-it\omega(p)}\hat{f}(-p)b(p)\right).$$

Hence the operator-valued distribution kernels of $\psi_\pm(t, \cdot)$ are given by

$$\psi_+(t, x) = \frac{1}{\sqrt{\ell}} \sum_{p \in \hat{T}_\ell} \frac{v(p)}{\sqrt{2\omega(p)}} \left(e^{it\omega(p)-ipx} b(p)^* + e^{-it\omega(p)+ipx} b(p) \right),$$

$$\psi_-(t, x) = -\frac{i}{\sqrt{\ell}} \sum_{p \in \hat{T}_\ell} \frac{v(-p)}{\sqrt{2\omega(p)}} \left(e^{it\omega(p)-ipx} b(p)^* - e^{-it\omega(p)+ipx} b(p) \right).$$

The CAR (5.13) is symbolically written as

$$\{\psi_\alpha(t, x), \psi_\beta(t, y)\} = \delta_{\alpha\beta} \delta(x - y), \quad \alpha, \beta = \pm. \tag{5.15}$$

As in the case of the boson Hamiltonian H_b (cf. [22, Proposition 9.12]), one can show that

$$H_f = \sum_{p \in \hat{T}_\ell} \omega(p) b(p)^* b(p) \quad \text{on Dom}(H_f).$$

Let $\psi_\pm(x) := \psi_\pm(0, x)$. Then one can define operator-valued distribution kernels $\xi_1(x), \xi_2(x)$ by

$$\xi_1(x) := \frac{1}{\sqrt{2}} (\psi_+(x) + \psi_-(x)),$$

$$\xi_2(x) := \frac{1}{\sqrt{2}} (\psi_+(x) - \psi_-(x)).$$

It follows from (5.15) that $\{\xi_1(x), \xi_2(x)\}$ also obeys the CAR:

$$\{\xi_a(x), \xi_b(y)\} = \delta_{ab} \delta(x - y), \quad a, b = 1, 2.$$

Explicitly, $\xi_a(x)$ has the following form as an operator-valued distribution kernel:

$$\xi_1(x) = \frac{1}{\sqrt{\ell}} \sum_{p \in \hat{T}_\ell} \frac{1}{\sqrt{2\omega(p)}} \left(\tau(p)^* b(p)^* e^{-ipx} + \tau(p) b(p) e^{ipx} \right),$$

$$\xi_2(x) = \frac{1}{\sqrt{\ell}} \sum_{p \in \hat{T}_\ell} \frac{1}{\sqrt{2\omega(p)}} \left(\tau(p) b(p)^* e^{-ipx} + \tau(p)^* b(p) e^{ipx} \right),$$

where

$$\tau(p) := \frac{1}{\sqrt{2}} (v(p) + iv(-p)), \quad p \in \hat{T}_\ell.$$

For each $f \in L^2(T_\ell)$, we define

$$\xi_1(f) := \sum_{p \in \hat{T}_\ell} \frac{1}{\sqrt{2\omega(p)}} \left(\tau(p)^* \hat{f}(p) b(p)^* + \tau(p) \hat{f}(-p) b(p) \right),$$

$$\xi_2(f) := \sum_{p \in \hat{T}_\ell} \frac{1}{\sqrt{2\omega(p)}} \left(\tau(p) \hat{f}(p) b(p)^* + \tau(p)^* \hat{f}(-p) b(p) \right),$$

which are bounded linear operators on $\mathscr{F}_f(\mathscr{H}_\ell)$, and write

$$\int_{T_\ell} \xi_j(x) f(x) dx := \xi_j(f), \quad j = 1, 2. \tag{5.16}$$

5.2   The $N = 1$ Wess–Zumino Model on M_ℓ

We are now ready to construct the Wess–Zumino models on the cylinder M_ℓ in a mathematically rigorous way. We refer the reader to [40] for a survey of these models as well as aspects related to them.

There are two kinds of Wess–Zumino models which are distinguished by the degree N of supersymmetry: $N = 1$ and $N = 2$. We first consider the $N = 1$ Wess–Zumino model on M_ℓ. The model describes an interaction of a relativistic neutral Bose field with a Majorana field. Hence a Hilbert space of state vectors for the $N = 1$ Wess–Zumino model can be taken to be the boson–fermion Fock space

$$\mathscr{F}_{bf}(\mathscr{H}_\ell) := \mathscr{F}_b(\mathscr{H}_\ell) \otimes \mathscr{F}_f(\mathscr{H}_\ell)$$

over $(\mathscr{H}_\ell, \mathscr{H}_\ell)$.

5.2.1   The Free Case

As a first step, we construct a *free* $N = 1$ Wess–Zumino model on M_ℓ. The Hamiltonian of it is defined by

$$H_0 := H_b \otimes I + I \otimes H_f,$$

where H_b and H_f are defined by (5.7) and (5.14) respectively.

To define a supercharge, let S_τ be the multiplication operator on $\mathscr{H}_\ell$ by the function $\tau(-p)^*$:

$$\mathrm{Dom}(S_\tau) = \left\{ u \in \mathscr{H}_\ell \mid \sum_{p \in \hat{T}_\ell} |\tau(-p)^* u(p)|^2 < \infty \right\},$$

$$(S_\tau u)(p) := \tau(-p)^* u(p), \quad p \in \hat{T}_\ell, \ u \in \mathrm{Dom}(S_\tau)$$

and denote by d_{S_τ} the boson–fermion Fock space exterior differential operator associated with S_τ (see (4.15)). Then, applying (4.7) and (4.9) to the present context, one can show that, for all $\Psi \in \mathscr{F}_{\mathrm{b,fin}}(\mathrm{Dom}(S_\tau)) \hat{\otimes} \mathscr{F}_{\mathrm{f,fin}}(\mathscr{H}_\ell)$,

$$d_{S_\tau} \Psi = \sum_{p \in \hat{T}_\ell} \tau(-p)^* b(p)^* a(p) \Psi.$$

The supercharge in the free $N = 1$ Wess–Zumino model is defined by the Dirac operator

$$Q_0 := Q_{S_\tau} = d_{S_\tau}^* + d_{S_\tau}.$$

Since

$$S_\tau^* S_\tau = \omega = S_\tau S_\tau^*,$$

we have by (4.41)

$$Q_0^2 = d\Gamma_{\mathrm{b}}(\omega) \otimes I + I \otimes d\Gamma_{\mathrm{f}}(\omega) = H_0.$$

Hence, letting Γ_ℓ be the state-sign operator on $\mathscr{F}_{\mathrm{bf}}(\mathscr{H}_\ell)$, we see that $(\mathscr{F}_{\mathrm{bf}}(\mathscr{H}_\ell), \Gamma_\ell, Q_0, H_0)$ is an SQM. This is a mathematically rigorous form of the $N = 1$ free Wess–Zumino model.

5.2.2 The Interacting Case

To define an interaction of the neutral quantum Klein–Gordon field and the Majorana field, it is more convenient to use the Q-space representation of the quantum Klein–Gordon field.

Let

$$L_{\mathrm{real}}^2(T_\ell) := \{ f \in L^2(T_\ell) \mid f = f^* \},$$

the real Hilbert space consisting of real elements in $L^2(T_\ell)$. Then, by (5.4), we have for all $f, g \in L_{\mathrm{real}}^2(T_\ell)$

$$\langle \Omega_{\mathscr{H}_\ell}, \phi(f)\phi(g)\Omega_{\mathscr{H}_\ell} \rangle = \frac{1}{2} \left\langle \frac{\hat{f}}{\sqrt{\omega}}, \frac{\hat{g}}{\sqrt{\omega}} \right\rangle_{\mathscr{H}_\ell} = \frac{1}{2} \langle h_\ell^{-1/2} f, h_\ell^{-1/2} g \rangle_{L^2(T_\ell)}, \quad (5.17)$$

where h_ℓ is defined by (5.2). Note that, for all $f \in L^2_{real}(T_\ell)$, $h_\ell^{-1/2} f$ is in $L^2_{real}(T_\ell)$.

Formula (5.17) suggests that $\phi(f)$ can be realized as the Gaussian random process indexed by the real Hilbert space

$$\mathscr{R}_\ell := \text{the completion of } L^2_{real}(T_\ell) \text{ with respect to the inner product}$$

$$\langle h_\ell^{-1/2} f, h_\ell^{-1/2} g \rangle_{L^2(T_\ell)} \ (f, g \in L^2_{real}(T_\ell)).$$

We denote by $\{\varphi(f) | f \in \mathscr{R}_\ell\}$ the Gaussian random process indexed by $\mathscr{R}_\ell$ and by $(Q_\ell, \Sigma_\ell, \mu_\ell)$ the underlying probability measure space. Then, by Theorem 3.1, there exists a unique unitary operator U_ℓ from $\mathscr{F}_b(\mathscr{R}_\ell)$ to $L^2(Q_\ell, d\mu_\ell)$ such that $U_\ell \Omega_{\mathscr{R}_\ell} = 1$ and, for all $n \geq 1$, $f_1, \ldots, f_n \in L^2_{real}(T_\ell)$,

$$U_\ell A_{\mathscr{R}_\ell}(f_1)^* \cdots A_{\mathscr{R}_\ell}(f_n)^* \Omega_{\mathscr{R}_\ell} = 2^{n/2} : \varphi(f_1) \cdots \varphi(f_n) : .$$

Remark 5.1 It is easy to see that, for all $\gamma > 2$, $h_\ell^{-(\gamma-1)}$ is trace class. Hence, applying the theory in Sect. 4.12.1 to the case where $\mathfrak{h} = L^2_{real}(T_\ell)$ and $h = h_\ell$, one can take, as Q_ℓ and $\varphi(f)$ ($f \in L^2_{real}(T_\ell)$), the following:

$$Q_\ell = \mathfrak{h}_{-\gamma} \ (\gamma > 2), \quad \varphi(f) = \langle \varphi, f \rangle, \quad \varphi \in \mathfrak{h}_{-\gamma}.$$

One of the advantages of this choice for Q_ℓ is in that Q_ℓ is a real Hilbert space and hence that infinite-dimensional differential calculus on Q_ℓ may be employed for further analysis.

It is easy to see that $h_\ell^{1/2}$ can be extended to a unitary operator from $L^2(T_\ell)$ to $\mathscr{R}_\ell$. We denote the extension by $\tilde{h}_\ell^{1/2}$. Then

$$J_\ell := \tilde{h}_\ell^{1/2} \mathscr{F}_\ell^{-1}$$

is a unitary operator from $\mathscr{H}_\ell$ to $\mathscr{R}_\ell$. Hence, by the theory in Sect. 3.4, the operator

$$V_\ell := U_\ell \Gamma_b(J_\ell) \tag{5.18}$$

is a unitary operator from $\mathscr{F}_b(\mathscr{H}_\ell)$ to $L^2(Q_\ell, d\mu_\ell)$, satisfying that $V_\ell \Omega_{\mathscr{H}_\ell} = 1$ and, for all $n \geq 1$, $f_1, \ldots, f_n \in L^2_{real}(T_\ell)$,

$$V_\ell a(\omega^{-1/2} \hat{f}_1)^* \cdots a(\omega^{-1/2} \hat{f}_n)^* \Omega_{\mathscr{H}_\ell} = 2^{n/2} : \varphi(f_1) \cdots \varphi(f_n) : .$$

In particular, we have

$$V_\ell \phi(f) V_\ell^{-1} = \varphi(f), \quad f \in L^2_{real}(T_\ell).$$

The subspace

$$\mathscr{W}_\ell := \text{span}\,\{1, :\varphi(f_1)\cdots\varphi(f_n): |\, n \geq 1, f_1, \ldots, f_n \in L^2_{\text{real}}(T_\ell) \cap D(h_\ell)\}$$

is dense in $L^2(Q_\ell, d\mu_\ell)$. It is easy to see that, for all $\Psi, \Phi \in \mathscr{W}_\ell$, the mapping $f \in \mathscr{D}_{\text{real}}(T_\ell) \mapsto \langle \Psi, \varphi(f)\Phi \rangle_{L^2(Q_\ell, d\mu_\ell)}$ is continuous. Hence $\langle \Psi, \varphi(\cdot)\Phi \rangle_{L^2(Q_\ell, d\mu_\ell)} \in \mathscr{D}'(T_\ell)$. In this sense, φ is a random distribution on T_ℓ. As in the case of ordinary distributions, we symbolically write

$$\varphi(f) = \int_{T_\ell} \varphi(x) f(x) dx$$

and call the symbol $\varphi(x)$ the distribution kernel of the random distribution φ.

In what follows, we use $L^2(Q_\ell, d\mu_\ell)$ as a Q-space representation of $\mathscr{F}_b(\mathscr{H}_\ell)$ and work with $L^2(Q_\ell, d\mu_\ell)$.

Let $\chi \in \mathscr{S}(\mathbb{R})$, the Schwartz space of rapidly decreasing C^∞-functions on $\mathbb{R}$, satisfying (i) $\int_{\mathbb{R}} \chi(x)dx = 1$; (ii) $\chi(x) = \chi(-x)$, $x \in \mathbb{R}$; (iii) the Fourier transform $\hat{\chi}$ of χ is non-negative; (iv) $\hat{\chi}(p) = 0$ for $|p| \geq 1$ and $\hat{\chi}(p) > 0$ for $|p| < 1/2$. For each $\kappa > 0$, let

$$\chi_\kappa(x) := \kappa \sum_{k \in \mathbb{Z}} \chi(\kappa(x - k\ell)), \quad x \in \mathbb{R}.$$

Then it follows that χ_κ is in $C^\infty(\mathbb{R})$ and periodic with period ℓ. Hence χ_κ can be regarded as an element of $\mathscr{D}(T_\ell)$. It is easy to show that, for all $f \in \mathscr{D}(T_\ell)$ and $x \in T_\ell$, $\lim_{\kappa \to \infty} \int_{T_\ell} \chi_\kappa(x - y)f(y)dy = f(x) = \delta_x(f)$. Namely, $\chi_\kappa(x - \cdot)$ converges to the delta-distribution δ_x at x in $\mathscr{D}'(T_\ell)$. Noting this property, one defines the time-zero quantum field $\varphi_\kappa(x)$ ($x \in T_\ell$) with ultraviolet cutoff κ in the Q-space representation by

$$\varphi_\kappa(x) := \varphi(\chi_\kappa(x - \cdot)),$$

which is mathematically meaningful as a random variable. By (3.2), we have

$$\int_{Q_\ell} :\varphi_\kappa(x)^n:^2 d\mu_\ell = \frac{n!}{2^n} \|\chi_\kappa(x - \cdot)\|^{2n}_{\mathscr{R}_\ell}.$$

Hence

$$\int_{T_\ell} dx \int_{Q_\ell} :\varphi_\kappa(x)^n:^2 d\mu_\ell = \frac{n!}{2^n} \|\chi_\kappa\|^{2n}_{\mathscr{R}_\ell} \ell < \infty.$$

Therefore, by the Fubini theorem, $\int_{T_\ell} :\varphi_\kappa(x)^n:^2 dx < \infty$ (μ_ℓ-a.e.). Hence it follows that, for all polynomials $P(x)$ of $x \in \mathbb{R}$,

$$\int_{T_\ell} |:P(\varphi_\kappa(x):|^2 dx < \infty \quad (\mu_\ell\text{-a.e.}). \tag{5.19}$$

Let W be a polynomial of the form

$$W(x) = V(x) - \frac{1}{2}mx^2, \quad x \in \mathbb{R}$$

with $V(x)$ being a real polynomial of degree $\deg V \geq 2$. Then

$$Q_{I,\kappa} := \int_{T_\ell} \xi_1(x) : W'(\varphi_\kappa(x)) : dx$$

—see(5.16)—is well-defined as a bounded linear operator on $\mathscr{F}_f(\mathscr{H}_\ell)$ (μ_ℓ-a.e.), i.e., $Q_{I,\kappa}$ is a $\mathcal{B}(\mathscr{F}_f(\mathscr{H}_\ell))$-valued function on (Q_ℓ, μ_ℓ). The supercharge of the interacting $N = 1$ Wess–Zumino model with ultraviolet cutoff κ is defined by

$$Q_\kappa := Q_0 + Q_{I,\kappa}.$$

It is easy to see that Q_κ is densely defined with $\text{Dom}(Q_\kappa) \supset \mathscr{W}_\ell$ and Q_κ is a symmetric operator.

For μ_ℓ-a.e. $q \in Q_\ell$, one can define $Y_\kappa(q) \in \mathscr{H}_\ell$ by

$$Y_\kappa(q)(p) := \frac{1}{\sqrt{\ell}} \int_{T_\ell} : W'(\varphi_\kappa(x)(q)) : e^{-ipx} dx, \quad p \in \hat{T}_\ell,$$

the discrete Fourier transform of the function $T_\ell \ni x \mapsto : W'(\varphi_\kappa(x)(q)) :$. It follows from the unitarity of the discrete Fourier transform that

$$\|Y_\kappa(q)\|_{\mathscr{H}_\ell}^2 = \int_{T_\ell} : W'(\varphi_\kappa(x)(q)) :^2 dx.$$

Note that

$$\frac{|\tau(p)|^2}{2\omega(p)} = 1.$$

Hence, μ_ℓ-a.e. $q \in Q_\ell$, $\tau^* Y_\kappa(q)/\sqrt{2\omega}$ is an element of $\mathscr{H}_\ell$. Therefore one can define a mapping $F_\kappa^{(1)} : Q_\ell \to \mathscr{H}_\ell$ by

$$F_\kappa^{(1)}(q) := \frac{\tau^*}{\sqrt{2\omega}} Y_\kappa(q), \quad \mu_\ell\text{-a.e. } q \in Q_\ell.$$

It is easy to see that $Q_{I,\kappa} = b(F_\kappa^{(1)}) + b(F_\kappa^{(1)})^*$ on $\mathscr{W}_\ell$. Hence we have

$$Q_\kappa = Q_{S_\tau}(F_\kappa^{(1)}) \quad \text{on } \mathcal{W}_\ell,$$

where $Q_{S_\tau}(F_\kappa^{(1)})$ is the operator $Q_S(F)$ (see (4.46)) with $S = S_\tau$ and $F = F_\kappa^{(1)}$. Thus the abstract SQM $(\mathfrak{F}, \Gamma_{\mathrm{BF}}, Q_S(F), H_S(F))$ yields, as a concrete realization, the interacting $N = 1$ Wess–Zumino model with ultraviolet cutoff.

In the paper [44], the following results are reported: (i) Q_κ is essentially self-adjoint (we denote its closure by the same symbol); (ii) $Q_{\kappa,+} := Q_\kappa \upharpoonright \mathrm{Dom}(Q_\kappa) \cap L^2(Q_\ell, d\mu_\ell; \mathcal{F}_{\mathrm{f},+}(\mathcal{H}_\ell))$ is Fredholm and ind $Q_{\kappa,+}$ is constant in κ. In particular, ind $Q_{\kappa,+} = $ ind $Q_{0,+}$; (iii). ind $Q_{\kappa,+} = \varepsilon[(\deg V + 1) \bmod 2]$, where $\varepsilon = \pm 1$ is the sign of the highest degree coefficient of V. It is shown that the index formula continues to hold also in the limit $\kappa \to \infty$ (the removal of the ultraviolet cutoff).

5.3 The $N = 2$ Wess–Zumino Model on M_ℓ

We next consider the $N = 2$ Wess–Zumino model on M_ℓ. This is a supersymmetric quantum field model which describes an interaction of a charged Bose field and the Dirac field.[2] We first recall these fields.

The Hilbert space for a charged particle on T_ℓ with spin 0 is taken, in the momentum representation, to be

$$\mathcal{K}_\ell := \mathcal{H}_\ell \oplus \mathcal{H}_\ell,$$

where $\mathcal{H}_\ell \oplus \{0\}$ (resp. $\{0\} \oplus \mathcal{H}_\ell$) describes the Hilbert space of state vectors of the particle (resp. anti-particle). Let $\mathcal{F}_{\mathrm{b}}(\mathcal{K}_\ell)$ be the boson Fock space over $\mathcal{K}_\ell$ and denote by $a(u, v)$ the boson annihilation operator with test vector $(u, v) \in \mathcal{K}_\ell$ on $\mathcal{F}_{\mathrm{b}}(\mathcal{K}_\ell)$. The annihilation operator $a_+(u)$ (resp. $a_-(u)$) $(u \in \mathcal{H}_\ell)$ for the particle (resp. anti-particle) is given by

$$a_+(u) := a(u, 0), \quad a_-(u) := a(0, u).$$

It follows from (2.9) and (2.10) that, for all $u, v \in \mathcal{H}_\ell$,

$$[a_\pm(u), a_\pm(v)^*] = \langle u, v \rangle,$$
$$[a_+(u), a_-(v)^\#] = 0, \quad [a_+(u)^\#, a_-(v)] = 0$$

on $\mathcal{F}_{\mathrm{b},0}(\mathcal{K}_\ell)$. The free charged Bose field is defined by

$$\phi_{\mathrm{c}}(t, f) := \frac{1}{\sqrt{2}} \left\{ a_+ \left(e^{it\omega} \frac{\hat{f}}{\sqrt{\omega}} \right)^* + a_- \left(e^{it\omega} \frac{\widehat{f^*}}{\sqrt{\omega}} \right) \right\}, \quad t \in \mathbb{R}, f \in L^2(T_\ell).$$

---

[2] A quantum mechanical version of the model is studied in a mathematically rigorous way in [7, 24, 27, 41].

Note that $\phi_c(t, f)$ is not symmetric even if $f = f^*$. Hence it is not neutral. As in the case of the neutral quantum scalar field $\phi(t, f)$, one can show that, for all $f \in \mathrm{Dom}(\Delta_\ell)$ and $\Psi \in \mathscr{F}_{b,0}(\mathscr{K}_\ell)$, the correspondence $\mathbb{R} \ni t \mapsto \phi_c(t, f)\Psi$ is twice strongly differentiable and the following equation holds:

$$\frac{d^2}{dt^2}\phi_c(t, f)\Psi + \phi_c(t, (-\Delta_\ell + m^2)f)\Psi = 0,$$

where d/dt means strong differentiation with respect to t. Thus the operator-valued functional $(t, f) \in \mathbb{R} \times \mathrm{Dom}(\Delta_\ell) \mapsto \phi_c(t, f)$ satisfies the free Klein–Gordon equation on the subspace $\mathscr{F}_{b,0}(\mathscr{K}_\ell)$.

The canonical conjugate momentum operator of $\phi_c(t, f)$ is defined by

$$\pi_c(t, f) := \frac{i}{\sqrt{2}}\left\{a_-(\sqrt{\omega}e^{it\omega}\hat{f})^* - a_+(\sqrt{\omega}e^{it\omega}\widehat{f^*})\right\} \tag{5.20}$$

for $f \in \mathrm{Dom}(h_\ell^{1/2})$. We have

$$\pi_c(t, f) = \frac{\partial \phi_c(t, f^*)^*}{\partial t} \quad \text{on } \mathscr{F}_{b,0}(\mathscr{K}_\ell).$$

It is easy to see that the following CCR are satisfied: for all $f_1, f_2 \in L^2(T_\ell)$ and $g_1, g_2 \in \mathrm{Dom}(h_\ell^{1/2})$,

$$[\phi_c(t, f_1), \pi_c(t, g_1)] = i \int_{T_\ell} f_1(x)g_1(x)dx, \tag{5.21}$$

$$[\phi_c(t, f_1), \pi_c(t, g_1)^*] = 0,$$

$$[\phi_c(t, f_1), \phi_c(t, f_2)^\#] = 0, \quad [\pi_c(t, g_1), \pi_c(t, g_2)^\#] = 0.$$

Let

$$H_c := d\Gamma_b(\omega \oplus \omega).$$

Then we have by (2.14)

$$\phi_c(t, f) = e^{itH_c}\phi_c(f)e^{-itH_c}, \quad \pi_c(t, g) = e^{itH_c}\pi_c(g)e^{-itH_c}$$

for all $t \in \mathbb{R}$, $f \in L^2(T_\ell)$, $g \in \mathrm{Dom}(h_\ell^{1/2})$, where

$$\phi_c(f) := \phi_c(0, f) = \frac{1}{\sqrt{2}} \left\{ a_+ \left(\frac{\hat{f}}{\sqrt{\omega}} \right)^* + a_- \left(\frac{\widehat{f^*}}{\sqrt{\omega}} \right) \right\},$$

$$\pi_c(f) := \pi_c(0, f) = \frac{i}{\sqrt{2}} \left\{ a_-(\sqrt{\omega}\hat{f})^* - a_+(\widehat{\sqrt{\omega}f^*}) \right\},$$

the time-zero field of $\phi_c(t, f)$ and of $\pi_c(t, g)$ respectively. Thus H_c is interpreted as the Hamiltonian of the charged Bose field under consideration.

For later use, here we write down the operator-valued distribution kernel of $\phi_c(f)$ and of $\pi_c(f)$:

$$\phi_c(x) = \frac{1}{\sqrt{\ell}} \sum_{p \in \hat{T}_\ell} \frac{1}{\sqrt{2\omega(p)}} \left(a_+(p)^* + a_-(-p) \right) e^{-ipx},$$

$$\pi_c(x) = \frac{i}{\sqrt{\ell}} \sum_{p \in \hat{T}_\ell} \frac{\sqrt{\omega(p)}}{\sqrt{2}} \left(a_-(p)^* - a_+(-p) \right) e^{-ipx},$$

where $a_\pm(p) := a_\pm(\delta_p)$.

We next define the free quantum Dirac field on T_ℓ. The Hilbert space of state vectors for the free quantum Dirac field is taken to be $\mathscr{F}_f(\mathscr{K}_\ell)$, the fermion Fock space over $\mathscr{K}_\ell$. We denote by $b(u, v)$ the fermion annihilation operator with test vector $(u, v) \in \mathscr{K}_\ell$ on $\mathscr{F}_f(\mathscr{K}_\ell)$. The annihilation operator $b_+(u)$ (resp. $b_-(u)$) ($u \in \mathscr{H}_\ell$) for the Dirac particle (resp. anti-Dirac particle) is given by

$$b_+(u) := b(u, 0), \quad b_-(u) := b(0, u).$$

It follows from (2.17) and (2.18) that, for all $u, v \in \mathscr{H}_\ell$,

$$\{b_\pm(u), b_\pm(v)^*\} = \langle u, v \rangle,$$

$$\{b_+(u), b_-(v)^\#\} = 0, \quad \{b_+(u)^\#, b_-(v)\} = 0.$$

In the $N = 2$ Wess–Zumino model on M_ℓ, it is suitable to use the following representation of the gamma matrices γ^0 and γ^1:

$$\gamma^0 = \begin{pmatrix} 0 & -1 \\ -1 & 0 \end{pmatrix}, \quad \gamma^1 = \begin{pmatrix} 0 & -1 \\ 1 & 0 \end{pmatrix}.$$

In this representation, we have

$$\alpha^1 = \begin{pmatrix} -1 & 0 \\ 0 & 1 \end{pmatrix}$$

and hence

$$\hat{h}_D(p) = \begin{pmatrix} -p & -m \\ -m & p \end{pmatrix}.$$

It follows that the eigenvalues of $\hat{h}_D(p)$ are $\pm\omega(p)$ and

$$w_+(p) := \frac{1}{\sqrt{2\omega(p)}}\begin{pmatrix} v(-p) \\ -v(p) \end{pmatrix}, \quad w_-(p) := \frac{1}{\sqrt{2\omega(p)}}\begin{pmatrix} v(p) \\ v(-p) \end{pmatrix}$$

are normalized eigenvectors of $\hat{h}_D(p)$ with eigenvalue $\omega(p)$ and $-\omega(p)$ respectively.
Let

$$b_\pm(p) := b_\pm(\delta_p).$$

Then one can show that the object $\psi(t, x) := (\psi_1(t, x), \psi_2(t, x))$ with the operator-valued distribution kernels

$$\psi_1(t, x) := \frac{1}{\sqrt{\ell}} \sum_{p\in\hat{T}_\ell} \frac{v(-p)}{\sqrt{2\omega(p)}} \left\{ b_-(p)^* e^{it\omega(p)-ipx} + b_+(p)e^{-it\omega(p)+ipx} \right\},$$

$$\psi_2(t, x) := \frac{1}{\sqrt{\ell}} \sum_{p\in\hat{T}_\ell} \frac{v(p)}{\sqrt{2\omega(p)}} \left\{ b_-(p)^* e^{it\omega(p)-ipx} - b_+(p)e^{-it\omega(p)+ipx} \right\},$$

satisfies (5.8) as an equation of operator-valued distributions and that the anti-commutation relations hold in the sense of operator-valued distributions:

$$\{\psi_a(t, x), \psi_b(t, y)^*\} = \delta_{ab}\delta(x - y), \quad \{\psi_a(t, x), \psi_b(t, y)\} = 0, \quad a, b = 1, 2.$$

Hence $\psi(t, x)$ is a canonically quantized free Dirac field.[3]
    The time-zero Dirac field is given by

$$\psi_a(x) := \psi_a(0, x), \quad a = 1, 2.$$

Then, it follows from (2.23) that

$$\psi_a(t, x) = e^{itH_D}\psi_a(x)e^{-itH_D}, \quad a = 1, 2,$$

in the sense of operator-valued distributions, where

$$H_D := d\Gamma_f(\omega \oplus \omega).$$

Hence H_D is interpreted as the Hamiltonian of the free quantum Dirac field.
    Let

$$\sigma(p) := -\frac{i}{\sqrt{2}}v(-p), \quad p \in \hat{T}_\ell$$

and define an operator S_σ on $\mathcal{H}_\ell$ as follows:

---

[3] Of course, mathematically meaningful is the object $\psi_a(t, f)$ smeared with $f \in \mathscr{D}(T_\ell)$ $(a = 1, 2)$.

$$\text{Dom}(S_\sigma) := \{u \in \mathcal{K}_\ell | \sqrt{\omega}\, u \in \mathcal{K}_\ell\},$$

$$(S_\sigma u)(p) := \begin{pmatrix} -\sigma(-p) & \sigma(p) \\ \sigma(p) & \sigma(-p) \end{pmatrix} u(p), \quad p \in \hat{T}_\ell,\ u \in \text{Dom}(S_\sigma).$$

Then it is easy to see S_σ is a densely defined closed linear operator and

$$S_\sigma^* = -S_\sigma, \tag{5.22}$$

where we have used the fact $\sigma(p)^* = -\sigma(p)$, $p \in \hat{T}_\ell$. It follows that

$$S_\sigma^* S_\sigma = S_\sigma S_\sigma^* = \omega \oplus \omega. \tag{5.23}$$

Note that $\{(\delta_p, 0), (0, \delta_p) | p \in \hat{T}_\ell\}$ is a CONS of $\mathcal{K}_\ell$. Hence

$$d_{S_\sigma} = \sum_{p \in \hat{T}_\ell} \left\{ a(S_\sigma^*(\delta_p, 0)) \otimes b((\delta_p, 0)^* + a(S_\sigma^*(0, \delta_p)) \otimes b((0, \delta_p)^* \right\}$$

on $\mathscr{F}_{\text{fin}}(\text{Dom}(S_\sigma), \mathcal{K}_\ell)$. By (5.22), we have

$$S_\sigma^*(\delta_p, 0) = (\sigma(-p)\delta_p, -\sigma(p)\delta_p), \quad S_\sigma^*(0, \delta_p) = (-\sigma(p)\delta_p, -\sigma(-p)\delta_p).$$

By these formulas and the anti-linearity of $a(u)$ in $u \in \mathcal{K}_\ell$, we obtain

$$d_{S_\sigma} = \sum_{p \in \hat{T}_\ell} \{ (-\sigma(-p)a_+(p) + \sigma(p)a_-(p)) \otimes b_+(p)^*$$
$$+ (\sigma(p)a_+(p) + \sigma(-p)a_-(p)) \otimes b_-(p)^* \}$$

on $\mathscr{F}_{\text{fin}}(\text{Dom}(S_\sigma), \mathcal{K}_\ell)$. A supercharge of the free $N = 2$ Wess–Zumino model is defined by the Dirac operator

$$Q_{S_\sigma} := d_{S_\sigma} + d_{S_\sigma}^*.$$

By (4.41) and (5.23), we have

$$Q_{S_\sigma}^2 = H_c \otimes I + I \otimes H_{\text{D}}.$$

Hence

$$H_{\text{WZ}} := H_c \otimes I + I \otimes H_{\text{D}}$$

is the Hamiltonian of the free $N = 2$ Wess–Zumino model.

We next consider the interacting $N = 2$ Wess–Zumino model. In the same way as in the case of the charged quantum scalar field on $\mathbb{R}^d$ (see [22, §10.16]), one can show that there exists a unitary operator X_ℓ from $\mathscr{F}_b(\mathscr{K}_\ell)$ to $\mathscr{F}_b(\mathscr{H}_\ell) \otimes \mathscr{F}_b(\mathscr{H}_\ell)$ such that $X_\ell \Omega_{\mathscr{K}_\ell} = \Omega_{\mathscr{H}_\ell} \otimes \Omega_{\mathscr{H}_\ell}$ and

$$X_\ell \overline{\phi_c(f)} X_\ell^{-1} = \frac{1}{\sqrt{2}} \overline{(\phi(f) \otimes I + i\, I \otimes \phi(f))}, \quad f \in L^2(T_\ell),$$

$$X_\ell H_c X_\ell^{-1} = H_b \otimes I + I \otimes H_b,$$

where $\phi(f)$ and H_b are defined by (5.4) and (5.7) respectively. Let V_ℓ be defined by (5.18). Then $\tilde{V}_\ell := V_\ell \otimes V_\ell$ is a unitary operator from $\mathscr{F}_b(\mathscr{H}_\ell) \otimes \mathscr{F}_b(\mathscr{H}_\ell)$ to $L^2(Q_\ell, d\mu_\ell) \otimes L^2(Q_\ell, d\mu_\ell) \cong L^2(Q_\ell \times Q_\ell, d(\mu_\ell \otimes \mu_\ell))$ such that $\tilde{V}_\ell \Omega_{\mathscr{H}_\ell} \otimes \Omega_{\mathscr{H}_\ell} = 1$ and, for all $f \in L^2_{\mathrm{real}}(T_\ell)$,

$$\tilde{V}_\ell(\phi(f) \otimes I)\tilde{V}_\ell^{-1} = \varphi_1(f), \quad \tilde{V}_\ell(I \otimes \phi(f))\tilde{V}_\ell^{-1} = \varphi_2(f),$$

where

$$\varphi_1(f)(q_1, q_2) := \varphi(q_1), \quad \varphi_2(q_1, q_2) := \varphi(f))(q_2), \quad \mu_\ell \otimes \mu_\ell\text{-a.e. } (q_1, q_2) \in Q_\ell \times Q_\ell.$$

Hence, letting $\Upsilon_\ell := \tilde{V}_\ell X_\ell$, we obtain that $\Upsilon_\ell \Omega_{\mathscr{K}_\ell} = 1$ and

$$\Upsilon \overline{\phi_c(f)} \Upsilon_\ell^{-1} = \frac{1}{\sqrt{2}}(\varphi_1(f) + i\varphi_2(f)), \quad f \in L^2_{\mathrm{real}}(T_\ell)$$

Thus $\overline{\phi_c(f)}$ is realized as a complex Gaussian random process. We call $\{\varphi_1(f) + i\varphi_2(f) | f \in L^2_{\mathrm{real}}(T_\ell)\}$ the Q-space representation of $\{\phi_c(f) | f \in L^2_{\mathrm{real}}(T_\ell)\}$. In what follows, we work with this representation.

For each $\kappa > 0$, we define

$$\Phi_\kappa(x) := \frac{1}{\sqrt{2}}\{\varphi_1(\chi_\kappa(x - \cdot)) + i\varphi_2(\chi_\kappa(x - \cdot))\}.$$

Let $U(z)$ be a polynomial of complex variable $z \in \mathbb{C}$ with degree $\deg U \geq 2$ and set

$$P(z) := U(z) - \frac{m}{2}z^2, \quad z \in \mathbb{C}.$$

As in the case of (5.19), one can show that

$$\int_{T_\ell} |P'(\Phi_\kappa(x))|^2 dx < \infty \quad (\mu_\ell \otimes \mu_\ell\text{-a.e.}).$$

Hence

$$Q_{\mathrm{I}}(\kappa) := -\frac{i}{\sqrt{2}} \int_{T_\ell} (\psi_1(x) P'(\Phi_\kappa(x)) + \psi_2(x) P'(\Phi_\kappa(x))^*) dx$$

$$+ \frac{i}{\sqrt{2}} \int_{T_\ell} (\psi_1(x)^* P'(\Phi_\kappa(x))^* + \psi_2(x)^* P'(\Phi_\kappa(x))) dx$$

is well-defined as a bounded linear operator on $\mathscr{F}_{\mathrm{f}}(\mathscr{H}_\ell)$ ($\mu_\ell \otimes \mu_\ell$-a.e.), i.e., $Q_{\mathrm{I}}(\kappa)$ is a $\mathfrak{B}(\mathscr{F}_{\mathrm{f}}(\mathscr{H}_\ell))$-valued function on $(Q_\ell \times Q_\ell, \mu_\ell \otimes \mu_\ell)$. A supercharge of the interacting $N = 2$ Wess–Zumino model with ultraviolet cutoff κ is given by

$$Q(\kappa) := Q_{S_\sigma} + Q_{\mathrm{I}}(\kappa),$$

an operator on $L^2(Q_\ell \times Q_\ell, d(\mu_\ell \otimes \mu_\ell); \mathscr{F}_{\mathrm{f}}(\mathscr{H}_\ell))$, where Q_{S_σ} is that in the Q-space representation (i.e., strictly writing, it is $\Upsilon Q_{S_\sigma} \Upsilon^{-1}$). It is easy to see that $Q(\kappa)$ is a symmetric operator.

For $\mu_\ell \otimes \mu_\ell$-a.e. $q \in Q_\ell \times Q_\ell$, one can define $Z_\kappa(q) \in \mathscr{H}_\ell$ by

$$Z_\kappa(q)(p) := \frac{1}{\sqrt{\ell}} \int_{T_\ell} P'(\Phi_\kappa(x)(q)) e^{-ipx} dx, \quad p \in \hat{T}_\ell.$$

It follows that

$$\|Z_\kappa(q)\|_{\mathscr{H}_\ell}^2 = \int_{T_\ell} |P'(\Phi_\kappa(x)(q))|^2 dx.$$

It is easy to see that $\sup_{p \in \hat{T}_\ell} |v(\pm p)/\sqrt{\omega(p)}| < \infty$. Hence one can define a mapping $F_\kappa^{(2)} : Q_\ell \times Q_\ell \to \mathscr{H}_\ell$ by

$$F_\kappa^{(2)}(q)(p) := \begin{pmatrix} \frac{i}{2\sqrt{\omega(p)}} (v(-p) Z_\kappa(-p)^* - v(p) Z_\kappa(p)) \\ -\frac{i}{2\sqrt{\omega(p)}} (v(-p) Z_\kappa(p) + v(p) Z_\kappa(-p)^*) \end{pmatrix}, \quad \mu_\ell \otimes \mu_\ell\text{-a.e. } q, \ p \in \hat{T}_\ell.$$

We have

$$\|F_\kappa^{(2)}(q)\|_{\mathscr{H}_\ell}^2 = \int_{T_\ell} |P'(\Phi_\kappa(x))|^2 dx.$$

The mapping $F_\kappa^{(2)}$ is in fact defined so that

$$Q_{\mathrm{I}}(\kappa) = b(F_\kappa^{(2)}) + b(F_\kappa^{(2)})^* \tag{5.24}$$

on the subspace

span $\{1, \varphi_1(f_1) \cdots \varphi_1(f_n)\varphi_2(g_1) \cdots \varphi_2(g_r)$

$|n, r \geq 0, n + r \geq 1, f_1, \ldots, f_n, g_1, \ldots, g_r \in L^2_{\text{real}}(T_\ell)\} \hat{\otimes} \mathscr{F}_{\text{f,fin}}(\mathscr{H}_\ell)$.

We redefine $Q_{\text{I}}(\kappa)$ by (5.24) and $Q(\kappa)$ by

$$Q(\kappa) := Q_{S_\sigma}(F_\kappa^{(2)}),$$

where $Q_{S_\sigma}(F_\kappa^{(2)})$ is the operator $Q_S(F)$ (see (4.46)) with $S = S_\sigma$ and $F = F_\kappa^{(2)}$. Thus the abstract SQM $(\mathfrak{F}, \Gamma_{\text{BF}}, Q_S(F), H_S(F))$ yields, as a concrete realization, the interacting $N = 2$ Wess–Zumino model with ultraviolet cutoff. In the paper [42, 43] (cf. also [39]), the following results are reported: (i) $Q(\kappa)$ is essentially self-adjoint (we denote its closure by the same symbol); (ii) $Q_+(\kappa) := Q(\kappa) \restriction \text{Dom}(Q(\kappa)) \cap L^2(Q_\ell \times Q_\ell, d(\mu_\ell \otimes \mu_\ell); \mathscr{F}_{\text{f},+}(\mathscr{H}_\ell))$ is Fredholm and ind $Q_+(\kappa)$ is constant in κ. In particular, ind $Q_+(\kappa) = $ ind $Q_+(0)$; (iii) ind $Q_+(\kappa) = \deg V - 1$. It is shown that the index formula continues to hold also in the limit $\kappa \to \infty$.

5.4   Other Models

There are supersymmetric quantum field models other than the Wess–Zumino models, to which the mathematical framework presented in Chap. 4 can be applied. Below is a list of them:

(i) *A model of a non-relativistic Fermi field interacting with a non-relativistic Bose field*. The *free Hamiltonian* of the model is of the form

$$d\Gamma_{\text{b}}(-\Delta_n + m^2) \otimes I + I \otimes d\Gamma_{\text{f}}(-\Delta_n + m^2)$$

on $\mathscr{F}_{\text{b}}(L^2(\mathbb{R}^n)) \otimes \mathscr{F}_{\text{f}}(L^2(\mathbb{R}^n))$ (see Remark 4.3). The model is associated with the so-called Parisi–Wu stochastic quantization [54]. See [4] for more details.

(ii) *A model of a non-relativistic Fermi field interacting with a gauge field* [70]. This model is related to the Floer theory [28].

(iii) *A model obtained as a supersymmetric extension   of a quantum scalar field model* [10].

(iv) *The Wess–Zumino–Witten model* [46, 47].

(v) *A model of a Bose field interacting with a Fermi field on the d-dimensional lattice* $\mathbb{Z}^d$ [52].

Appendix A
Self-adjoint Extensions of a Symmetric Operator Matrix

Let $\mathcal{H}_1$ and $\mathcal{H}_2$ be Hilbert spaces and

$$\mathcal{H} := \mathcal{H}_1 \oplus \mathcal{H}_2 = \{\Psi = (\Psi_1, \Psi_2)|\Psi_1 \in \mathcal{H}_1, \Psi_2 \in \mathcal{H}_2\}, \qquad (A.1)$$

the direct sum Hilbert space of $\mathcal{H}_1$ and $\mathcal{H}_2$. Let L be a linear operator on $\mathcal{H}$ such that

$$\mathrm{Dom}(L) = (\mathrm{Dom}(L) \cap \mathcal{H}_1) \oplus (\mathrm{Dom}(L) \cap \mathcal{H}_2). \qquad (A.2)$$

Then, for $a, b = 1, 2$, one can define a linear operator L_{ab} from $\mathcal{H}_b$ to $\mathcal{H}_a$ as follows:

$$\mathrm{Dom}(L_{ab}) := \mathrm{Dom}(L) \cap \mathcal{H}_b,$$
$$L_{a1}\Psi_1 := (L(\Psi_1, 0))_a, \quad L_{a2}\Psi_2 := (L(0, \Psi_2))_a, \quad \Psi_b \in \mathrm{Dom}(L_{ab}).$$

Then we have
$$L\Psi = (L_{11}\Psi_1 + L_{12}\Psi_2, L_{21}\Psi_1 + L_{22}\Psi_2), \quad \Psi \in \mathrm{Dom}(L).$$

In this sense, we write

$$L = \begin{pmatrix} L_{11} & L_{12} \\ L_{21} & L_{22} \end{pmatrix}. \qquad (A.3)$$

This representation is called the operator matrix representation of L with respect to (A.1).

If L is bounded with $\mathrm{Dom}(L) = \mathcal{H}$, then (A.2) is satisfied and hence L has always the operator matrix representation (A.3) with L_{ab} being bounded with $\mathrm{Dom}(L_{ab}) = \mathcal{H}_b$.

© The Author(s), under exclusive license to Springer Nature Singapore Pte Ltd. 2022
A. Arai, *Infinite-Dimensional Dirac Operators and Supersymmetric Quantum Fields*,
SpringerBriefs in Mathematical Physics,
https://doi.org/10.1007/978-981-19-5678-2

Conversely, suppose that, for $a, b = 1, 2$, a linear operator L_{ab} from $\mathscr{H}_b$ to $\mathscr{H}_a$ is given. Then these operators define a linear operator L on $\mathscr{H}$ by (A.3) with $\mathrm{Dom}(L) = (\mathrm{Dom}(L_{11}) \cap \mathrm{Dom}(L_{21})) \oplus (\mathrm{Dom}(L_{12}) \cap \mathrm{Dom}(L_{22}))$.

In this appendix, we consider only the case where L is anti-diagonal, i.e., $L_{11} = 0$ and $L_{22} = 0$.[1] In this case, L takes the form

$$A := \begin{pmatrix} 0 & T \\ S & 0 \end{pmatrix}, \tag{A.4}$$

with T (resp. S) being a linear operator from $\mathscr{H}_2$ (resp. $\mathscr{H}_1$) to $\mathscr{H}_1$ (resp. $\mathscr{H}_2$). We have

$$\mathrm{Dom}(A) = \mathrm{Dom}(S) \oplus \mathrm{Dom}(T). \tag{A.5}$$

Remark A.1 There is another simple case of L, i.e., the case where $L_{12} = 0$ and $L_{21} = 0$ so that

$$L = L_{\mathrm{D}} := \begin{pmatrix} L_{11} & 0 \\ 0 & L_{22} \end{pmatrix}.$$

In this case, L is said to be diagonal. It is easy to see that $L_{\mathrm{D}} = L_{11} \oplus L_{22}$, the direct sum operator of L_{11} and L_{22}.

Some basic properties of the anti-diagonal operator matrix A are summarized in the following lemma:

Lemma A.1 Let A be the operator matrix given by (A.4).

(i) The operator A is closed if and only if S and T are closed.

(ii) The operator A is closable if and only if S and T are closable. In that case, the closure $\bar{A}$ of A is given as follows:

$$\bar{A} = \begin{pmatrix} 0 & \bar{T} \\ \bar{S} & 0 \end{pmatrix}. \tag{A.6}$$

(iii) The operator A is densely defined if and only S and T are densely defined. In that case, the adjoint A^* of A is given as follows:

$$A^* = \begin{pmatrix} 0 & S^* \\ T^* & 0 \end{pmatrix}. \tag{A.7}$$

Proof (i) Suppose that A is closed. Let $\Psi_n \in \mathrm{Dom}(S)$ ($n \in \mathbb{N}$) be a sequence such that $\lim_{n\to\infty} \Psi_n = \Psi \in \mathscr{H}_1$ and $\lim_{n\to\infty} S\Psi_n = \Phi \in \mathscr{H}_2$. Then, $(\Psi_n, 0) \in \mathrm{Dom}(A)$, $\lim_{n\to\infty} (\Psi_n, 0) = (\Psi, 0)$ and $\lim_{n\to\infty} A(\Psi_n, 0) = (0, \Phi)$. Hence, by the closedness of A, $(\Psi, 0) \in \mathrm{Dom}(A)$ and $A(\Psi, 0) = (0, \Phi)$. This means that $\Psi \in \mathrm{Dom}(S)$ and

---

[1] For more general cases, see [22, Appendix B].

$S\Psi = \Phi$. Therefore S is closed. Similarly, for any sequence $\Psi'_n \in \mathrm{Dom}(T)$ $(n \in \mathbb{N})$ such that $\lim_{n\to\infty} \Psi'_n = \Psi' \in \mathscr{H}_2$ and $\lim_{n\to\infty} T\Psi'_n = \Phi' \in \mathscr{H}_1$, one can prove that $\Psi' \in \mathrm{Dom}(T)$ and $T\Psi' = \Phi'$. Hence T is closed.

Conversely, suppose that S and T are closed. Let $\Psi_n = (\Psi_{n1}, \Psi_{n2}) \in \mathrm{Dom}(A)$ $(n \in \mathbb{N})$ be a sequence such that $\lim_{n\to\infty} \Psi_n = \Psi \in \mathscr{H}$ and $\lim_{n\to\infty} A\Psi_n = \Phi \in \mathscr{H}$. Then $\lim_{n\to\infty} \Psi_{na} = \Psi_a$ $(a = 1, 2)$ and $\lim_{n\to\infty} S\Psi_{n1} = \Phi_2$, $\lim_{n\to\infty} T\Psi_{n2} = \Phi_1$. Since S and T are closed, it follows that $\Psi_1 \in \mathrm{Dom}(S)$, $\Psi_2 \in \mathrm{Dom}(T)$ and $S\Psi_1 = \Phi_2$, $T\Psi_2 = \Phi_1$. Hence S and T are closed.

(ii) Suppose that A is closable. Let $\Psi_n \in \mathrm{Dom}(S)$ $(n \in \mathbb{N})$ be a sequence such that $\lim_{n\to\infty} \Psi_n = 0$ and $\lim_{n\to\infty} S\Psi_n = \Phi \in \mathscr{H}_2$. Then $\lim_{n\to\infty} (\Psi_n, 0) = (0, 0)$ and $\lim_{n\to\infty} A(\Psi_n, 0) = (0, \Phi)$. Since A is closable, it follows that $(0, \Phi) = (0.0)$. Hence $\Phi = 0$. Therefore S is closable. Similarly, one can show that T is closable.

Conversely, suppose that S and T are closable. Then one can define the following operator matrix:

$$\tilde{A} := \begin{pmatrix} 0 & \bar{T} \\ \bar{S} & 0 \end{pmatrix}.$$

By (i), $\tilde{A}$ is a closed operator. It is obvious that $A \subset \tilde{A}$ (i.e., $\tilde{A}$ is an extension of A). Hence A is closable and $\bar{A} \subset \tilde{A}$.

To prove the converse relation $\tilde{A} \subset \bar{A}$, let $\Psi \in \mathrm{Dom}(\tilde{A}) = \mathrm{Dom}(\bar{S}) \oplus \mathrm{Dom}(\bar{T})$. Then there exists a sequence $\{\Psi_n\}_{n=1}^{\infty}$ with $\Psi_{n1} \in \mathrm{Dom}(S)$ and $\Psi_{n2} \in \mathrm{Dom}(T)$ such that $\lim_{n\to\infty} \Psi_{n1} = \Psi_1$, $\lim_{n\to\infty} S\Psi_{n1} = \bar{S}\Psi_1$ and $\lim_{n\to\infty} \Psi_{n2} = \Psi_2$, $\lim_{n\to\infty} T\Psi_{n2} = \bar{T}\Psi_2$. This implies that $\Psi_n \in \mathrm{Dom}(A)$ and $\lim_{n\to\infty} \Psi_n = \Psi$, $\lim_{n\to\infty} A\Psi_n = (\bar{T}\Psi_2, \bar{S}\Psi_1)$. Hence $\Psi \in \mathrm{Dom}(\bar{A})$ and $\bar{A}\Psi = (\bar{T}\Psi_2, \bar{S}\Psi_1) = \tilde{A}\Psi$. Therefore $\tilde{A} \subset \bar{A}$. Thus we obtain $\tilde{A} = \bar{A}$, i.e., (A.6) holds.

(iii) By (A.5), A is densely defined if and only if S and T are densely defined. Suppose that A is densely defined and let

$$A' := \begin{pmatrix} 0 & S^* \\ T^* & 0 \end{pmatrix}.$$

Then $\mathrm{Dom}(A') = \mathrm{Dom}(T^*) \oplus \mathrm{Dom}(S^*)$ and, for all $\Phi \in \mathrm{Dom}(A)$, $\Psi \in \mathrm{Dom}(A')$,

$$\langle \Psi, A\Phi \rangle_{\mathscr{H}} = \langle \Psi_1, T\Phi_2 \rangle_{\mathscr{H}_1} + \langle \Psi_2, S\Phi_1 \rangle_{\mathscr{H}_2} = \langle A'\Psi, \Phi \rangle_{\mathscr{H}}.$$

Hence $\Psi \in \mathrm{Dom}(A^*)$ and $A^*\Psi = A'\Psi$. This means that $A' \subset A^*$.

Conversely, let $\Psi \in \mathrm{Dom}(A^*)$. Then, for all $\Phi \in \mathrm{Dom}(A)$, we have $\langle A\Phi, \Psi \rangle_{\mathscr{H}} = \langle \Phi, A^*\Psi \rangle_{\mathscr{H}}$. Let $A^*\Psi = (\eta_1, \eta_2)$. Then $\langle T\Phi_2, \Psi_1 \rangle_{\mathscr{H}_1} + \langle S\Phi_1, \Psi_2 \rangle_{\mathscr{H}_2} = \langle \Phi_1, \eta_1 \rangle_{\mathscr{H}_1} + \langle \Phi_2, \eta_2 \rangle_{\mathscr{H}_2}$. Take $\Phi_2 = 0$. Then $\langle S\Phi_1, \Psi_2 \rangle_{\mathscr{H}_2} = \langle \Phi_1, \eta_1 \rangle_{\mathscr{H}_1}$. Since $\Phi_1 \in \mathrm{Dom}(S)$ is arbitrary, it follows that $\Psi_2 \in \mathrm{Dom}(S^*)$ and $S^*\Psi_2 = \eta_1$. We next take $\Phi_1 = 0$. Then $\langle T\Phi_2, \Psi_1 \rangle_{\mathscr{H}_1} = \langle \Phi_2, \eta_2 \rangle_{\mathscr{H}_2}$. Hence $\Psi_1 \in \mathrm{Dom}(T^*)$ and $T^*\Psi_1 = \eta_2$. Therefore $\Psi \in \mathrm{Dom}(T^*) \oplus \mathrm{Dom}(S^*)$ and $A^*\Psi = (S^*\Psi_2, T^*\Psi_1)$. This means that $\Psi \in \mathrm{Dom}(A')$ and $A'\Psi = A^*\Psi$, i.e., $A^* \subset A'$. Thus $A^* = A'$, i.e., (A.7) holds. $\square$

The symmetricity and the self-adjointness of A given by (A.4) are characterized as follows:

Theorem A.1 *Let A be the operator matrix given by (A.4).*

(i) *The operator A is symmetric if and only if S and T are densely defined and $T \subset S^*$, $S \subset T^*$.*

(ii) *The operator A is self-adjoint if and only if S is closed and $T = S^*$.*

Proof (i) Suppose that A is symmetric. Then A is densely defined and $A \subset A^*$. Hence, by (A.5), S and T are densely defined and (A.7) imply that $T \subset S^*$ and $S \subset T^*$.

Conversely, suppose that S and T are densely defined and $T \subset S^*$ and $S \subset T^*$. Then, by (A.5), A is densely defined. By (A.7), we have $A \subset A^*$. Hence A is symmetric.

(ii) The operator A is self-adjoint if and only if $A^* = A$. By (A.7), this is equivalent to that $T = S^*$ and $S = T^*$. Since the adjoint of a densely defined linear operator is closed, it follows that $T = S^*$ and $S = T^*$ if and only if S is closed and $T = S^*$ (recall that, for a densely defined closable linear operator C, $C^{**} = \bar{C}$). Thus the assertion is proved. □

Theorem A.2 *Let A be the operator matrix given by (A.4). Suppose that A is symmetric. Then S and T are densely defined closable and the operators*

$$A_1 := \begin{pmatrix} 0 & S^* \\ \bar{S} & 0 \end{pmatrix}, \quad A_2 := \begin{pmatrix} 0 & \bar{T} \\ T^* & 0 \end{pmatrix}, \tag{A.8}$$

are self-adjoint extensions of A, i.e., A_1 and A_2 are self-adjoint operators satisfying $A \subset A_1$ and $A \subset A_2$.

Proof By Theorem A.1(i), S and T are densely defined and $S \subset T^*$, $T \subset S^* \cdots (*)$. These relations imply that T^* and S^* are densely defined. Hence T and S are closable and $T^{**} = \bar{T}$, $S^{**} = \bar{S}$. Hence it follows from Theorem A.1(ii) that A_1 and A_2 are self-adjoint. Relations $(*)$ imply that $A \subset A_1$ and $A \subset A_2$. Hence A_1 and A_2 are self-adjoint extensions of A. □

Remark A.2 It is easy to see that $A_1 \neq A_2$ if and only if A is not essentially self-adjoint.

Finally, we describe the rule of the product of two anti-diagonal operator–matrices:

Lemma A.2 *Let A be given by (A.4) and*

$$B = \begin{pmatrix} 0 & V \\ U & 0 \end{pmatrix}$$

with U (resp. V) being a linear operator from $\mathcal{H}_1$ (resp. $\mathcal{H}_2$) to $\mathcal{H}_2$ (resp. $\mathcal{H}_1$). Then operator equality

$$AB = TU \oplus SV = \begin{pmatrix} TU & 0 \\ 0 & SV \end{pmatrix} \qquad (A.9)$$

holds.

Proof An easy exercise.   □

Appendix B
Construction of an Infinite-Dimensional Gaussian Measure on a Path Space

In this appendix, we give an outline on the construction of Gaussian measure μ_β in Lemma 4.20 in a more general setting.

Let $\beta > 0$ and

$$k_n := \frac{2\pi n}{\beta}, \quad n \in \mathbb{Z}.$$

Lemma B.1 *Let $\lambda > 0$ and $0 < |t| < \beta$ ($t \in \mathbb{R}$). Then the following formulas hold:*

$$\frac{1}{\beta} \sum_{n=-\infty}^{\infty} \frac{e^{itk_n}}{\lambda - ik_n} = \frac{e^{t\lambda}}{e^{\beta\lambda} - 1}, \quad t > 0, \tag{B.1}$$

$$\frac{1}{\beta} \sum_{n=-\infty}^{\infty} \frac{e^{itk_n}}{\lambda - ik_n} = \frac{e^{(\beta+t)\lambda}}{e^{\beta\lambda} - 1}, \quad t < 0. \tag{B.2}$$

Proof Equation (B.1) can be proved by applying the residue theorem to the meromorphic function $f(z) = e^{tz}(e^{\beta z} - 1)^{-1}(z - \lambda)^{-1}$ of complex variable z. Equation (B.2) follows from (B.1) with t replaced by $\beta + t > 0$ and the property $e^{i\beta k_n} = 1$, $n \in \mathbb{Z}$. □

Lemma B.2 *Let $\lambda > 0$ and $0 \le |t| \le \beta$ ($t \in \mathbb{R}$). Then*

$$\frac{e^{-|t|\lambda} + e^{-(\beta-|t|)\lambda}}{1 - e^{-\beta\lambda}} = \frac{2}{\beta} \sum_{n=-\infty}^{\infty} \frac{\lambda e^{itk_n}}{\lambda^2 + k_n^2}. \tag{B.3}$$

Proof This follows from (B.1) and (B.2). □

Let $\mathcal{H}$ be a real separable Hilbert space and $C([0, \beta]; \mathcal{H})$ be the space of $\mathcal{H}$-valued continuous functions on the interval $[0, \beta]$. We denote by $\mathcal{H}_{\mathbb{C}}$ the complexifi-

© The Author(s), under exclusive license to Springer Nature Singapore Pte Ltd. 2022   105
A. Arai, *Infinite-Dimensional Dirac Operators and Supersymmetric Quantum Fields*,
SpringerBriefs in Mathematical Physics,
https://doi.org/10.1007/978-981-19-5678-2

cation of $\mathscr{H}$. For each $F \in C([0, \beta]; \mathscr{H})$, we define the discrete Fourier transform $\hat{F} : \mathbb{Z} \to \mathscr{H}_{\mathbb{C}}$ by

$$\hat{F}(n) := \int_0^\beta \phi_n(t)^* F(t) dt, \quad \phi_n(t) := \frac{1}{\sqrt{\beta}} e^{itk_n}, \quad n \in \mathbb{Z}, t \in [0, \beta],$$

where the integral is taken in the sense of a strong Riemann integral in $\mathscr{H}_{\mathbb{C}}$. It is well-known that $\{\phi_n\}_{n=-\infty}^\infty$ is a CONS of $L^2([0, \beta])$. Hence, for all CONS $\{e_m\}_{m=1}^\infty$ of $\mathscr{H}_{\mathbb{C}}$, $\{\phi_n \otimes e_m\}_{n \in \mathbb{Z}, m \in \mathbb{N}}$ with $(\phi_n \otimes e_m)(t) := \phi_n(t) e_m$ is a CONS of $L^2([0, \beta]; \mathscr{H}_{\mathbb{C}})$ (the Hilbert space of $\mathscr{H}_{\mathbb{C}}$-valued measurable functions on $[0, \beta]$). Using this fact, one can show that $\sum_{n=-\infty}^\infty \|\hat{F}(n)\|_{\mathscr{H}_{\mathbb{C}}}^2 < \infty$, satisfying

$$\sum_{n=-\infty}^\infty \|\hat{F}(n)\|_{\mathscr{H}_{\mathbb{C}}}^2 = \int_0^\beta \|F(t)\|_{\mathscr{H}}^2 dt.$$

Let A be a strictly positive self-adjoint operator on $\mathscr{H}$ with $A \geq a > 0$. Then, by the functional calculus of self-adjoint operators, for all $f \in \mathscr{H}_{\mathbb{C}}$ and $n \in \mathbb{Z}$, we have $\|A(A^2 + k_n^2)^{-1} f\|_{\mathscr{H}_{\mathbb{C}}} \leq \|f\|_{\mathscr{H}_{\mathbb{C}}}/a$. Hence, using the Schwarz inequality, one can show that, for all $F, G \in C([0, \beta]; \mathscr{H})$, $\sum_{n=-\infty}^\infty |\langle \hat{F}(n), A(A^2 + k_n^2)^{-1} \hat{G}(n) \rangle_{\mathscr{H}_{\mathbb{C}}}| < \infty$. Therefore one can define a sesquilinear form $\langle F, G \rangle_{A,-1}$ by

$$\langle F, G \rangle_{A,-1} := 2 \sum_{n=-\infty}^\infty \langle \hat{F}(n), A(A^2 + k_n^2)^{-1} \hat{G}(n) \rangle_{\mathscr{H}_{\mathbb{C}}}, \quad F, G \in C([0, \beta]; \mathscr{H}).$$

Since $\ker(A^{1/2}(A^2 + k_n^2)^{-1/2}) = \{0\}$, it follows that $\langle , \rangle_{A,-1}$ is an inner product of $C([0, \beta]; \mathscr{H})$. We denote the completion of the real inner product space $(C([0, \beta]; \mathscr{H}), \langle , \rangle_{A,-1})$ by $\mathscr{W}$, which is a real separable Hilbert space.

For each $s \in [0, \beta]$, we denote by δ_s the delta distribution on $[0, \beta]$, i.e., $\delta_s \in \mathscr{D}'([0, \beta])$ such that $\delta_s(u) = u(s), u \in \mathscr{D}([0, \beta])$ (the space of infinitely differentiable periodic functions on $[0, \beta]$). The discrete Fourier transformation can be extended to $\delta_s \otimes f$ ($f \in \mathscr{H}$) with

$$\widehat{\delta_s \otimes f}(n) = \frac{1}{\sqrt{\beta}} e^{-isk_n} f.$$

Hence it follows that, for all $t \in [0, \beta]$ and $f \in \mathscr{H}$, $\delta_t \otimes f$ is in $\mathscr{W}$. Moreover, for all $f, g \in \mathscr{H}$ and $s, t \in [0, \beta]$,

$$\langle \delta_t \otimes f, \delta_s \otimes g \rangle_{\mathscr{W}} = \frac{2}{\beta} \sum_{n=-\infty}^{\infty} e^{i(t-s)k_n} \left\langle f, A(A^2 + k_n^2)^{-1} g \right\rangle_{\mathscr{H}}$$

$$= \frac{2}{\beta} \sum_{n=-\infty}^{\infty} \int_{(0,\infty)} e^{i(t-s)k_n} \frac{\lambda}{\lambda^2 + k_n^2} d \left\langle f, E_A(\lambda) g \right\rangle_{\mathscr{H}},$$

where E_A is the spectral measure of A. It is easy to see that one can interchange the integral $\int_{(0,\infty)} d \langle f, E_A(\lambda) g \rangle_{\mathscr{H}}$ and the summation $\sum_{n=-\infty}^{\infty}$. Then, using (B.3), we obtain

$$\langle \delta_t \otimes f, \delta_s \otimes g \rangle_{\mathscr{W}} = \left\langle f, (e^{-|t-s|A} + e^{-(\beta-|t-s|)A})(1 - e^{-\beta A})^{-1} g \right\rangle_{\mathscr{H}}. \qquad \text{(B.4)}$$

We now consider the Gaussian random process $\phi_{\mathscr{W}}$ indexed by $\mathscr{W}$ so that

$$\int_{Q_{\mathscr{W}}} e^{i\phi_{\mathscr{W}}(F)} d\mu_{\mathscr{W}} = e^{-\|F\|_{\mathscr{W}}^2/4}, \quad \int_{Q_{\mathscr{W}}} \phi_{\mathscr{W}}(F)\phi_{\mathscr{W}}(G) d\mu_{\mathscr{W}} = \frac{1}{2} \langle F, G \rangle_{\mathscr{W}}, \quad F, G \in \mathscr{W},$$

where $(Q_{\mathscr{W}}, \mu_{\mathscr{W}})$ is the underlying probability measure space. Hence, by (B.4),

$$\int_{Q_{\mathscr{W}}} \phi_{\mathscr{W}}(\delta_t \otimes f)\phi_{\mathscr{W}}(\delta_s \otimes g) d\mu_{\mathscr{W}}$$

$$= \frac{1}{2} \left\langle f, (e^{-|t-s|A} + e^{-(\beta-|t-s|)A})(1 - e^{-\beta A})^{-1} g \right\rangle_{\mathscr{H}}, \quad s, t \in [0, \beta], f, g \in \mathscr{H}. \qquad \text{(B.5)}$$

In what follows, we assume that, for some constant $\gamma > 1$, $A^{-(\gamma-1)}$ is trace class. Then $A^{-\gamma/2}$ is Hilbert–Schmidt. The domain $\mathrm{Dom}(A^{\gamma/2})$ is a real Hilbert space with inner product $\langle f, g \rangle_\gamma := \left\langle A^{\gamma/2} f, A^{\gamma/2} g \right\rangle_{\mathscr{H}}$, $f, g \in \mathrm{Dom}(A^{\gamma/2})$. We denote this Hilbert space by $\mathscr{H}_\gamma$. The sesquilinear form $\langle \, , \, \rangle_{-\gamma} : \mathscr{H} \times \mathscr{H} \to \mathbb{R}$ defined by

$$\langle f, g \rangle_{-\gamma} := \left\langle A^{-\gamma/2} f, A^{-\gamma/2} g \right\rangle_{\mathscr{H}}, \quad f, g \in \mathscr{H},$$

is an inner product of $\mathscr{H}$. We denote the completion of the inner product space $(\mathscr{H}, \langle \, , \, \rangle_{-\gamma})$ by $\mathscr{H}_{-\gamma}$. It is shown that the dual space $\mathscr{H}_\gamma^*$ of $\mathscr{H}_\gamma$ is naturally isomorphic to $\mathscr{H}_\gamma$ by the natural bilinear form $\langle \phi, f \rangle$ such that $\langle \phi, f \rangle = \left\langle A^{-\gamma/2}\phi, A^{\gamma/2} f \right\rangle_{\mathscr{H}}$ for $\phi \in \mathscr{H}, f \in \mathscr{H}_\gamma$.

Since the embedding of $\mathscr{H}$ into $\mathscr{H}_{-\gamma}$ is Hilbert–Schmidt (nuclear), it follows from a general theorem (Minlos's theorem) that there exists a Gaussian measure μ_0 on $\mathscr{H}_{-\gamma}$ such that

$$\int_{\mathscr{H}_{-\gamma}} e^{i\langle \phi, f \rangle} d\mu_0 = e^{-\|f\|_{\mathscr{H}}^2/4}, \quad f \in \mathscr{H}_\gamma.$$

Since $A^{-\gamma}$ is trace class and positive, there exists a CONS $\{e_n\}_{n=1}^{\infty}$ of $\mathscr{H}$ and a sequence $\{\lambda_n\}_{n=1}^{\infty}$ of positive numbers such that $A e_n = \lambda_n e_n$ and $\sum_{n=1}^{\infty} \lambda_n^{-\gamma} < \infty$. Each $f \in \mathscr{H}$ is expanded as $f = \sum_{n=1}^{\infty} a_n e_n$, $a_n := \langle e_n, f \rangle_{\mathscr{H}}$. Then, for all $t \in [0, \beta]$,

$$\phi_{\mathscr{W}}(\delta_t \otimes f) = \sum_{n=1}^{\infty} a_n \phi_{\mathscr{W}}(\delta_t \otimes e_n)$$

in the topology of $L^2(\mathscr{W}, d\mu_{\mathscr{W}})$. For each $N \in \mathbb{N}$, we define

$$X_t^{(N)} = \sum_{n=1}^{N} \phi_{\mathscr{W}}(\delta_t \otimes e_n) e_n \in \mathscr{H}.$$

Then one can show that, for all $M, N \in \mathbb{N}$ with $M > N$,

$$\int_{Q_{\mathscr{W}}} \|X_t^{(N)} - X_t^{(M)}\|_{-\gamma}^2 d\mu_{\mathscr{W}} = \frac{1}{2} \sum_{n=N+1}^{M} \frac{1}{\lambda_n^{\gamma}} \coth \frac{\beta \lambda_n}{2} \to 0 \ (M, N \to \infty).$$

Hence the limit

$$X_t := \sum_{n=1}^{\infty} \phi_{\mathscr{W}}(\delta_t \otimes e_n) e_n$$

exists in $L^2(Q_{\mathscr{W}}, d\mu_{\mathscr{W}}; \mathscr{H}_{-\gamma})$ (the Hilbert space of $\mathscr{H}_{-\gamma}$-valued L^2-functions on $(Q_{\mathscr{W}}, \mu_{\mathscr{W}})$). Moreover, it is easy to see that $\langle X_t, f \rangle = \phi_{\mathscr{W}}(\delta_t \otimes f)$, $f \in \mathscr{H}_{\gamma}$. Hence $\{\langle X_t, f \rangle | t \in [0, \beta], f \in \mathscr{H}_{\gamma}\}$ is a family of jointly Gaussian random variables such that, for all $n \in \mathbb{N}$, $t_j \in [0, \beta]$ and $f_j \in \mathscr{H}_{\gamma}$,

$$\int_{\mathscr{W}} e^{i \sum_{j=1}^{n} \langle X_{t_j}, f_j \rangle} = e^{-\sum_{j,k=1}^{n} M_{jk}/4}, \quad M_{jk} := \int_{Q_{\mathscr{W}}} \langle X_{t_j}, f_j \rangle \langle X_{t_k}, f_k \rangle d\mu_{\mathscr{W}},$$

and (B.5) takes the form:

$$\int_{Q_{\mathscr{W}}} \langle X_t, f \rangle \langle X_s, g \rangle d\mu_{\mathscr{W}} = \frac{1}{2} \langle f, (e^{-|t-s|A} + e^{-(\beta - |t-s|)A})(1 - e^{-\beta A})^{-1} g \rangle_{\mathscr{H}},$$

$$s, t \in [0, \beta], \ f, g \in \mathscr{H}_{\gamma}. \tag{B.6}$$

Using the assumption that $A^{-(\gamma-1)}$ is trace class, one can show that, for all $s, t \in [0, \beta]$ and $n \in \mathbb{N}$,

$$\int_{Q_{\mathscr{W}}} \|X_t - X_s\|_{-\gamma}^{2n} d\mu_{\mathscr{W}} \leq C|t - s|^n$$

with $C > 0$ being a constant. Hence, by an application of Kolmogorov's lemma on regularity of stochastic process (see, e.g., [62, Theorem 5.1], [20, Corollary 2.23]), $\{X_t\}_{t \in [0,\beta]}$ has a continuous version (in fact, a Hölder continuous version). We denote by the continuous version of X_t by the same symbol so that, for a.e. $q \in Q_\mathcal{W}$, $X_t(q) \in \mathcal{H}_{-\gamma}$ is continuous in $t \in [0, \beta]$.

We now define a mapping $X : Q_\mathcal{W} \to C([0, \beta]; \mathcal{H}_{-\gamma})$ by

$$X(q)(t) := X_t(q), \quad \text{a.e. } q \in Q.$$

Let ν_β be the image measure by X so that, for all mappings $\eta : C([0, \beta]; \mathcal{H}_{-\gamma}) \to \mathbb{R}$,

$$\int_{Q_\mathcal{W}} \eta(X(q)) d\mu_\mathcal{W} = \int_{C([0,\beta];\mathcal{H}_{-\gamma})} \eta(\Phi) d\nu_\beta(\Phi).$$

In particular, taking η as $\eta(\Phi) := \langle \Phi(t), f \rangle \langle \Phi(s), g \rangle$, $\Phi \in C([0, \beta]; \mathcal{H}_{-\gamma})$, $f, g \in \mathcal{H}_\gamma$, we have by (B.6)

$$\int_{C([0,\beta];\mathcal{H}_{-\gamma})} \langle \Phi(t), f \rangle \langle \Phi(s), g \rangle \, d\nu_\beta(\Phi)$$

$$= \frac{1}{2} \langle f, (e^{-|t-s|A} + e^{-(\beta-|t-s|)A})(1 - e^{-\beta A})^{-1} g \rangle_\mathcal{H}, \, s, t \in [0, \beta], \, f, g \in \mathcal{H}_\gamma.$$

(B.7)

Thus we have shown that there exists a family $\{\langle \Phi(t), f \rangle \mid t \in [0, \beta], f \in \mathcal{H}_{-\gamma}\}$ of jointly Gaussian random variables on $C([0, \beta]; \mathcal{H}_{-\gamma})$ with an infinite-dimensional measure ν_β satisfying (B.7). Since $C([0, \beta]; \mathcal{H}_{-\gamma})$ is a path space with paths in $\mathcal{H}_{-\gamma}$, ν_β is a path space measure.

References

1. Albeverio, S., Kondrat'ev, Yu.G.: Supersymmetric Dirichlet operators. Ukrainian Math. J. **47**, 675–685 (1995)
2. Albeverio, S., Daletskii, A., Kondratiev, Yu.: Stochastic analysis on product manifolds: Dirichlet operators on differential forms. J. Funct. Anal. **176**, 280–316 (2000)
3. Albeverio, S., Daletskii, A., Lytvynov, E.: Laplace operators on differential forms over configuration spaces. J. Geom. Phys. **37**, 15–46 (2001)
4. Arai, A.: On a mathematical construction of supersymmetric quantum field theory associated with Parisi–Wu stochastic quantization. MPI-PAE/PTh 60/84 (Preprint), Max-Planck-Institut für Physik und Astrophysik (1984); unpublished
5. Arai, A.: Supersymmetry and singular perturbations. J. Funct. Anal. **60**, 378–393 (1985)
6. Arai, A.: A path integral representation of the index of Kähler-Dirac operators on an infinite dimensional manifold. J. Funct. Anal. **82**, 330–369 (1989)
7. Arai, A.: On the degeneracy in the ground state of the $N = 2$ Wess-Zumino supersymmetric quantum mechanics. J. Math. Phys. **30**, 2973–2977 (1989)
8. Arai, A.: A general class of infinite dimensional Dirac operators and related aspects. In: Koshi, S. (Ed.), Functional Analysis & Related Topics, pp. 85–98. World Scientific, Singapore (1991)
9. Arai, A.: A general class of infinite dimensional Dirac operators and path integral representation of their index. J. Funct. Anal. **105**, 342–408 (1992)
10. Arai, A.: Supersymmetric extension of quantum scalar field theories. In: Quantum and Noncommutative Analysis (Kyoto, 1992), pp. 73–90. Mathematical Physics Studies, vol. 16. Kluwer Academic Publishers, Dordrecht (1993)
11. Arai, A.: Dirac operators in boson–fermion Fock spaces and supersymmetric quantum field theory. Infinite-Dimensional Geometry in Physics (Karpacz, 1992). J. Geom. Phys. **11**, 465–490 (1993)
12. Arai, A.: On self-adjointness of Dirac operators in boson-fermion Fock spaces. Hokkaido Math. J. **23**, 319–353 (1994)
13. Arai, A.: Operator-theoretical analysis of a representation of a supersymmetry algebra in Hilbert space. J. Math. Phys. **36**, 613–621 (1995)
14. Arai, A.: Scaling limit of anticommuting self-adjoint operators and applications to Dirac operators. Integr. Equat. Oper. Th. **21**, 139–173 (1995)
15. Arai, A.: Trace formulas, a Golden-Thompson inequality and classical limit in boson Fock space. J. Funct. Anal. **136**, 510–546 (1996)

© The Author(s), under exclusive license to Springer Nature Singapore Pte Ltd. 2022
A. Arai, *Infinite-Dimensional Dirac Operators and Supersymmetric Quantum Fields*,
SpringerBriefs in Mathematical Physics,
https://doi.org/10.1007/978-981-19-5678-2

16. Arai, A.: Strong anticommutativity of Dirac operators on boson-fermion Fock spaces and representations of a supersymmetry algebra. Math. Nachr. **207**, 61–77 (1999)
17. Arai, A.: Fock Spaces and Quantum Fields I, II (in Japanese). Nippon Hyoronhsa, Tokyo (2000)
18. Arai, A.: Mathematical Principles of Quantum Phenomena (in Japanese). Asakura-Shoten, Tokyo (2006)
19. Arai, A.: Mathematical Principles of Quantum Statistical Mechanics (in Japanese). Kyoritsu-Schuppan, Tokyo (2008)
20. Arai, A.: Functional Integral Methods in Quantum Mathematical Physics (in Japanese). Kyoritsu-shuppan, Tokyo (2010)
21. Arai, A.: A special class of infinite dimensional Dirac operators on the abstract boson–fermion Fock space. J. Math. Art. ID 713690, 13 pp (2014)
22. Arai, A.: Analysis on Fock Spaces and Mathematical Theory of Quantum Fields. World Scientific, Singapore (2018)
23. Arai, A.: Inequivalent Representations of Canonical Commutation and Anti-Commutation Relations. Springer, Singapore (2020)
24. Arai, A., Hayashi, K.: Spectral analysis of a Dirac operator with a meromorphic potential. J. Math. Anal. Appl. **306**, 440–461 (2005)
25. Arai, A., Mitoma, I.: De Rham-Hodge-Kodaira decomposition in ∞-dimensions. Math. Ann. **291**, 51–73 (1991)
26. Arai, A., Mitoma, I.: Comparison and nuclearity of spaces of differential forms on topological vector spaces. J. Funct. Anal. **111**, 278–294 (1993)
27. Arai, A., Ogurisu, O.: Meromorphic $N = 2$ Wess-Zumino supersymmetric quantum mechanics. J. Math. Phys. **32**, 2427–2434 (1991)
28. Atiyah, M.: New invariants of 3- and 4-dimensional manifolds. The Mathematical Heritage of Hermann Weyl (Durham, NC, 1987), pp. 285–299. Proceedings of Symposia in Pure Mathematics, vol. 48. American Mathematical Society, Providence, RI (1988)
29. Bratteli, O., Robinson, D.W.: Operator Algebras and Quantum Statistical Mechanics, vol. 2. Springer, New York (1981)
30. Brüning, J., Lesch, M.: Hilbert complexes. J. Funct. Anal. **108**, 88–132 (1992)
31. Deift, P.A.: Applications of a commutation formula. Duke Math. J. **45**, 267–310 (1978)
32. Gel'fand, I.M., Vilenkin, N.Y.: Generalized Functions, vol. 4. Academic Press, New York (1964)
33. Gilkey, P.: Invariance Theory, the Heat Equation and the Atiyah-Singer Index Theorem, 2nd edn. CRC Press, New York (1995)
34. Glimm, J., Jaffe, A: Quantum Physics. Springer, New York (1981), Second Edition (1987)
35. Gross, L.: Potential theory on Hilbert space. J. Funct. Anal. **1**, 123–181 (1967)
36. Gross, L.: On the formula of Mathews and Salam. J. Funct. Anal. **25**, 162–209 (1977)
37. Hino, M.: Spectral properties of Laplacians on an abstract Wiener space with a weighted Wiener measure. J. Funct. Anal. **147**, 485–520 (1997)
38. Hiroshima, F., Lörinczi, J.: Feynman–Kac–Type Theorems and Gibbs Measures on Path Space, Volume 2: Applications in Rigorous Quantum Field Theory, 2nd edn. De Gruyter, Berlin/Boston (2020)
39. Jaffe, A., Lesniewski, A.: A priori estimates for the $N = 2$ Wess-Zumino model on a cylinder. Commun. Math. Phys. **114**, 553–575 (1988)
40. Jaffe, A., Lesniewski, A.: Supersymmetric quantum fields and infinite dimensional analysis. In: t Hooft, G., Jaffe, A., Mack, G., Mitter, P.K., Stora, R. (eds.) Nonperturbative Quantum Field Theory. Plenum Press, New York (1988)
41. Jaffe, A., Lesniewski, A., Lewenstein, M.: Ground state structure in supersymmetric quantum mechanics. Ann. Phys. (N.Y.) **178**, 313–329 (1987)
42. Jaffe, A., Lesniewski, A., Weitsman, J.: Index of a family of Dirac operators on loop space. Commun. Math. Phys. **112**, 75–88 (1987)
43. Jaffe, A., Lesniewski, A., Weitsman, J.: The two-dimensional, $N = 2$ Wess-Zumino model on a cylinder. Commun. Math. Phys. **114**, 147–165 (1988)

44. Jaffe, A., Lesniewski, A., Weitsman, J.: The loop space $S^1 \to \mathbb{R}$ and supersymmetric quantum fields. Ann. Phys. (NY) **183**, 337–351 (1988)
45. Kato, T.: Perturbation Theory for Linear Operators, 2nd edn. Springer, Berlin (1976)
46. Léandre, R.: Stochastic Wess-Zumino-Witten model over a symplectic manifold. J. Geom. Phys. **21**, 307–336 (1997)
47. Léandre, R.: Stochastic Wess–Zumino–Witten model for the measure of Kontsevitch. Seminar on Stochastic Analysis, Random Fields and Applications (Ascona 1996), pp. 231–247. Progress in Probability, vol. 45. Birkhäuser, Basel (1999)
48. Léandre, R.: Cover of the Brownian bridge and stochastic symplectic action. Rev. Math. Phys. **12**, 91–137 (2000)
49. Léandre, R.: A stochastic approach to the Euler-Poincaré characteristic of a quotient of a loop group. Rev. Math. Phys. **13**, 1307–1322 (2001)
50. Léandre, R., Roan, S.S.: A stochastic approach to the Euler-Poincaré number of the loop space of a developable orbifold. J. Geom. Phys. **16**, 71–98 (1995)
51. Lörinczi, J., Hiroshima, F., Betz, V.: Feynman–Kac–Type Theorems and Gibbs Measures on Path Space, Volume 1: Feynman–Kac–Type Formulae and Gibbs Measures, 2nd edn. De Gruyter, Berlin/Boston (2020)
52. Matte, O.: Supersymmetric Dirichlet operators, spectral gaps, and correlations. Ann. Henri Poincaré **7**, 731–780 (2006)
53. Miyao, T.: Strongly supercommuting self-adjoint operators. Integr. Equat. Oper. Th. **50**, 505–535 (2004)
54. Parisi, G., Wu, Y.S.: Perturbation theory without gauge fixing. Sci. Sinica **24**, 483–496 (1981)
55. Reed, M., Simon, B.: Methods of Modern Mathematical Physics I: Functional Analysis. Academic Press, New York (1972); Revised and enlarged edition (1980)
56. Reed, M., Simon, B.: Methods of Modern Mathematical Physics II: Fourier Analysis and Self-adjointness. Academic Press, New York (1975)
57. Reed, M., Simon, B.: Methods of Modern Mathematical Physics IV: Analysis of Operators. Academic Press, New York (1978)
58. Schwartz, L.: Méthodes Mathématiques Pour Les Science Physiques. Editions Scientifiques Hermann, Paris (1961)
59. Shigekawa, I.: Derivatives of Wiener functionals and absolute continuity of induced measures. J. Math. Kyoto Univ. **20–2**, 263–289 (1980)
60. Shigekawa, I.: De Rham-Hodge-Kodaira's decomposition on an abstract Wiener space. J. Math. Kyoto Univ. **26–2**, 191–202 (1986)
61. Simon, B.: The $P(\phi)_2$ Euclidean (Quantum) Field Theory. Princeton University Press, Princeton, NJ (1974)
62. Simon, B.: Functional Integration and Quantum Physics. Academic Press (1979)
63. Simon, B.: Trace Ideals and Their Applications, 2nd edn. American Mathematical Society, Providence, RI (2005)
64. Thaller, B.: The Dirac Equation. Springer, Berlin (1992)
65. Vasilescu, F.-H.: Anticommuting self-adjoint operators. Rev. Roum. Math. Pure Appl. **28**, 77–91 (1983)
66. Wess, J., Bagger, J.: Supersymmetry and Supergravity. Princeton University Press, Princeton (1983)
67. Witten, E.: Dynamical breaking of supersymmetry. Nucl. Phys. **B188**, 513–555 (1981)
68. Witten, E.: Constraints on supersymmetry breaking. Nucl. Phys. **B202**, 253–316 (1982)
69. Witten, E.: Supersymmetry and Morse theory. J. Diff. Geom. **17**, 661–692 (1982)
70. Witten, E.: Topological quantum field theory. Commun. Math. Phys. **117**, 353–386 (1988)

Index

Printed in the United States
by Baker & Taylor Publisher Services